11-5

Lessons from an Optical Illusion

It is the mark of the trained mind never to expect
more precision in the treatment of any subject
than the nature of that subject permits.

Aristotle, *The Nicomachean Ethics*

Lessons from an Optical Illusion

On Nature and Nurture, Knowledge and Values

Edward M. Hundert

Harvard University Press
Cambridge, Massachusetts
London, England
1995

Library of Congress Cataloging-in-Publication Data
Hundert, Edward M.
Lessons from an optical illusion : on nature and nurture,
knowledge and values / Edward M. Hundert.
p.   cm.
Includes bibliographical references and index.
ISBN 0-674-52540-X (alk. paper)
1. Knowledge, Theory of.   2. Values.   3. Nature and nurture.
I. Title.
BD161.H86   1995
155.2'34'01—dc20
94–37095
CIP

Dedicated with love to Mary,
my wonderful partner in this
and everything else

Preface

In a *Feminist Dictionary* published ten years ago, the entry under "Acknowledgments" reads in part: "Before feminism, that portion of a book where men thanked their wives for critically reading their manuscripts without asking for coauthorship." By this standard of input and recognition, I figure that *Lessons from an Optical Illusion* should include not only my wife but about two dozen other coauthors who critically read the manuscript and provided significant input. Indeed, if significant input were to become the standard, then the two dozen living coauthors would have to make room for at least as many other historical figures, whose seminal insights are what make a book like this possible. Let me provide a bit of background and thank some of these people by name.

About six years ago, I completed a book entitled *Philosophy, Psychiatry, and Neuroscience: Three Approaches to the Mind*. That project sought a synthetic, interdisciplinary approach to some of the basic philosophical problems of knowledge, combining insights from philosophy, psychiatry, psychology, neurobiology, and other related cognitive sciences. While it closed with a short appendix called "Cognitive and Moral Intersubjectivity," it did not otherwise address questions about values or morality, and the entire thing was such a dense, academic volume that—although a surprising number of people bought it—I doubt many of them ever actually finished it.

In order to extend that earlier line of thought to the realm of values and morality, this new book presents the earlier ideas in brief compass and then develops a parallel argument in the field of ethics, ultimately seeking to unify a comprehensive approach to both knowledge and values. A reading of the first book is not assumed herein, although it is recommended to those who, upon finishing this book—which was written to be finished—would like a more detailed account of the theory of knowledge.

One of the ironies of my professional life is the discovery that this more accessible, less academic book was actually the more challenging one to write. The five-year effort it took to produce a book that avoids technical jargon across multiple disciplines has taught me an important lesson—a lesson about how such jargon actually masks all sorts of conceptual confusions that I have tried to resolve herein by translating these concepts and conveying their meanings in a more straightforward way.

In this new project, I found myself faced with several perils in trying to build an interdisciplinary synthesis of ideas that might be enjoyed by any curious person. These perils include the risk of potentially insulting some scholars with a too-brief summary of a complex idea, defining terms that some think are obvious while avoiding some jargon others find indispensable, and never bringing each argument fully up to date with all the reviews of the latest claims and disputes and all the rebuttals to potential criticisms on each side—reviews and rebuttals that might be expected of a scholarly book within the confines of each separate discipline.

I fear I shall be guilty of all of these at times, and I of course bear responsibility for these deficits. But I have braved these perils in what I consider to be a good cause, namely, a book in which specialists and nonspecialists alike might find something of interest that challenges their perspectives. My guiding principle has been to incorporate my love of teaching not just into the ideas but into their exposition, so that the reader might be drawn with me into the process of interdisciplinary synthesis. In this spirit, Jean Piaget once made the observation that "real comprehension of a notion or a theory implies the reinvention of this theory by the subject." It would be a compliment to me as author to discover the reader reinventing this work in the course of reading it.

I am deeply indebted to the many people who have already involved themselves in the creation of this book. My thanks go first to several generous people for their critical reading of an early draft of the manuscript, including Thomas Benjamin, Larry Benowitz, Gerald Edelman, Joseph Gally, Alfred Margulies, Jonas Salk, Mark Sullivan, Paul Verschure, and especially Leston Havens, who has nurtured not only this project but my entire career. I would also like to thank a number of others who provided their significant input through discussions over an hour, over a weekend, or in some cases over the years, including Jeffrey Gilbert, Jerome Kagan, Barbara Maher, Brendan Maher, David Potter, Spencer Rice, Thomas Steindler, Frederick Will, and especially my philosophical mentor, Geoffrey Warnock.

Of course, Edelman, Salk, and Warnock are also three of the thinkers upon whose works this book is built. While some of the other primary sources are also living scientists, philosophers, and psychiatrists, such as Hubel and Wiesel, Modell, Putnam, and Rawls, most authors of these sources are long gone, from Plato and Aristotle, to Kant and Hume, to Freud and Winnicott, to Russell and Wittgenstein. These posthumous acknowledgments are actually the most important of all in a book that builds its argument by presenting and developing the ideas of these fascinating characters. Simply put, this book is their story—not mine. In a sense, the "new idea" here is the process by which their stories are woven together, the result being not so much new *particular* thoughts about the nature and nurture of knowledge and values, but a new *system* of thought about the nature and nurture of knowledge and values.

This book was written in a number of inspirational settings, including the magnificent libraries of Harvard Medical School, McLean Hospital, Wellesley College, and the Marine Biological Laboratories at Woods Hole. A considerable portion was also written at Supuko in Edinburgh, Scotland. I would like to thank all of these institutions for their hospitality and for sharing so generously of their resources.

I am also indebted to a number of individuals at Harvard Medical

School and McLean Hospital who have supported my academic career to ensure that time for a project like this could be added to the administrative, clinical, and teaching responsibilities through which I earn my keep. Thanks for this support go to Bruce Cohen, Joseph Coyle, Steven Mirin, Daniel Tosteson, and especially Daniel Federman, who, through all the years I have toiled on this book, continually sought creative ways to help me dedicate time to my scholarly interests.

Having listed so many people, it is hard to believe that even more have still not been mentioned, but these are my patients, whose acknowledgment will remain intentionally anonymous. My sincere gratitude also goes out to the many Harvard students and McLean psychiatry residents who have helped shape so much of what is in these pages by asking those insightful questions that cut to the heart of complex problems.

Thanks also go to my editor at Harvard University Press, Angela von der Lippe, for her faith in this project over many years. I would like to give special acknowledgment to Fiona McCrae, who provided a most useful editorial critique of the very first complete draft. Thanks also to my assistant, Kate Cox, who helped prepare the volume for publication.

A lifetime of gratitude goes to my generous, loving parents. I am truly blessed to have been the recipient of the nature and nurture of two such wonderful human beings.

Most of all, I want to thank my wife, Mary, for her never-ending love and support even when this project took me away from our family and friends. She is not just a coauthor of this book, but the coauthor of my life.

EMH

Illustration Credits

Charles Darwin: Photograph by Leonard Darwin. From *Transactions of the Shropshire Archeological and Natural History Society,* October 1884, Volume 8, Part I.

René Descartes: Portrait by David Beck. Reproduced courtesy of the Royal Swedish Academy of Sciences.

David Hume: Portrait by Allan Ramsay. Reproduced courtesy of the Scottish National Portrait Gallery.

Immanuel Kant: Portrait reproduced by permission of The Bettmann Archive.

Georg W. F. Hegel: Portrait reproduced by permission of The Bettmann Archive.

Bertrand Russell: Photograph reproduced by permission of The Bettmann Archive.

Hilary Putnam: Photograph reproduced courtesy of the Harvard University Office of News and Public Affairs.

Sigmund Freud: Photograph reproduced by permission of A. W. Freud et al., Mark Paterson and Associates.

Jean Piaget: Photograph reproduced by permission of The Bettmann Archive.

Wilder Penfield: Photograph reproduced courtesy of Charles P. Hodge.

David Hubel and Torsten Wiesel: Photograph reproduced courtesy of Harvard Medical School.

Gerald Edelman: Photograph reproduced by permission of Abe Frajndlich. Copyright © 1994 Abe Frajndlich.

Stephen Hawking: Photograph reproduced by permission of Manni Mason's Pictures.

Donald Winnicott: Photograph reproduced courtesy of Harvard University Press.

Jonas Salk: Photograph reproduced courtesy of the Salk Institute.

Sir Francis Galton: Photograph reproduced by permission of The Bettmann Archive.

Hemi-neglect drawings: From R. L. Straub and F. W. Black, *The Mental Status Examination in Neurology* (Philadelphia: F. A. Davis Company, 1993). Reproduced by permission of the publisher.

Contents

Part One

We must at all costs avoid over-simplification, which one might be tempted to call the occupational disease of philosophers if it were not their occupation.

J. L. Austin

1

The State of the Art

The relative influence of nature and nurture in determining who we are as human beings has been debated with vigor since ancient times. What could be more intriguing? "She certainly inherited your sense of humor," my friends tease me when my daughter manages to find something to laugh about in almost any human interaction. These friends—philosophers, psychiatrists, and genetics researchers among them—usually will not commit on whether they mean that her humor genes came from the personality-determining DNA of her father's chromosomes or whether they think this trait runs in our family in the same way the recipe does for how we cook the turkey at Thanksgiving.

This lighthearted side of the old debate takes on more urgency when a researcher claims that criminal behavior also runs in families. Virtually every aspect of the nature-nurture debate is rife with moral overtones. Even the courts have recently begun to entertain the argument that the defendant should be found innocent of his proven criminal behavior because he was so abused throughout his childhood. With an upbringing like his, how could we expect him to be a law-abiding citizen today? How can we in good conscience punish him *further* by convicting him for doing what any of us might have done had we received his cruel brand of "nurturing" (hardly worthy of the name)?

But consider for a moment the alternative explanation. Researchers suddenly announce that there is a gene for the particular type of criminal behavior committed by this fellow. It is a dominant gene that sits on the short arm of chromosome number six, let's say. Expert witnesses are brought in and demonstrate conclusively that the defendant inherited this unfortunate gene, presumably from his father (who spent most of his adult life in jail for similar crimes). We can almost hear the defense attorney. Saddled from birth with the tragic impact of his criminal gene, how could we expect him to be a law-abiding citizen today? How can we in good conscience punish him *further* by convicting him for doing what any of us might have done had we inherited this same gene?

How curious it is that the nurture argument (years of abuse) and the nature argument (an unfortunate gene) can each be used to absolve our man from moral responsibility for his actions. These seem to be our only two options, so from where can his culpability arise? Ironically, most of us believe that the only respectable answer to such classical either-or dilemmas as "nature or nurture" is inevitably "it's always some of both." Taking the comfortable position that he probably had some lousy genes *and* a rotten upbringing, we conclude: GUILTY AS CHARGED! Presumably, the matter gets complicated enough when nature and nurture mix together that—no longer able to point the finger at his parents' behavior or at the DNA in his chromosomes—we cannot find anywhere else to point than at the man himself.

The moral tension that permeates the nature-nurture debate can perhaps be best appreciated in its historical context when we recall the full title of Charles Darwin's 1859 publication: *On the Origin of Species by Means of Natural Selection or the Preservation of Favoured Races in the Struggle for Life.* The notion of favored races led in a very few years to the development of the eugenics movement. Indeed, the term "eugenics" was coined by Darwin's cousin, the explorer, anthropologist (and eugenicist) Sir Francis Galton. Ten short years after Darwin's book came out, Galton in turn published his *Hereditary Genius: An Inquiry into Its Laws and Consequences,* which is generally held to include the first formal articulation of the nature-nurture dichotomy. By studying the families of eminent scientists, judges, authors, musicians, military leaders, and theologians, Galton proved to his own satisfaction that the traits which lead to human "eminence" are heritable. As a relative of the great Charles Darwin, he presumably had some self-interest in this finding!

With his student Karl Pearson, Galton also developed the mathematical tools of regression analysis that we still use to tease apart correlations, and his 1876 article "The history of twins as a criterion of the relative powers of nature and nurture" firmly established the alliterative phrase for the contributions of genetics and environment that we still use today. Soon, other researchers began studying families at the other end of the social and intellectual spectrum, like the infamous Jukes family in New York. After the discovery that six of the Jukeses were in prison at the same time in just one county, a seven-generation review of about 750 Jukeses revealed that the family had cost the state over $1.5 million through crime, pauperism, and vice—and that was back in the nineteenth century when $1.5 million was a *lot* of money.

These questions had, of course, captured the imagination of people long before Darwin and Galton. As early as the thirteenth century, Frederick II, King of Germany, devised an "experiment" to see what language children would speak if he ordered their foster mothers to feed, bathe, and clothe them but never speak to or around them. His hypothesis was that they might speak Hebrew, which he thought was the oldest language (or perhaps Greek, Latin, or Arabic), or, if not, then the language of their biological parents. Unfortunately, virtually all of the children died, and he concluded instead that the loving words of a mother are needed for human survival. Three centuries later, the Mughal emperor Akbar (a descendant of Genghis Khan) reared children in isolation to discover whether their natural religion would be Hinduism or Christianity. This experiment also failed, producing only deaf mutes whose religious creed could not be determined.

Over the years of my academic life, I have always been intrigued by the many and varied ways nature and nurture issues have been handled by philosophers and theologians, biologists and sociologists, educators and politicians. But the practical implications of these intellectual issues always hit me squarely in the face when I care for patients who suffer from psychiatric illnesses. When the tragedy of a severe mental illness like schizophrenia strikes a teenage boy, his parents invariably come into my office not only with desperate hopes for a cure but also with some heart-rending questions that fall within the ancient nature-nurture debate. "Could this have been prevented?" they ask with guilt-ridden fears. "Should we have done something different when he was younger or is it some kind of inborn problem that was inevitable?"

The answers offered to an anxious family in a typical psychiatrist's office at any given time will reflect the nature-nurture ethos of the day. Although people have almost always, upon reflection, articulated the ever-respectable "some of both" position mentioned above, history reveals something of a pendulum swing, with each side of this debate periodically superseding the other. Thirty or so years ago, these parents would probably not get much of an answer from the doctor. The nurture side of the debate was in prominence, and all of his medical and psychiatric training would have taught a doctor back then that inadequate childrearing led to psychotic breakdown in the teenage years. In his reluctance to make the parents feel any more guilt, a doctor three decades ago would typically make some tangential comments and try to change the subject. Making people feel worse than they already do was not considered part of the physician's role, and what could he say to these people whose cold and inconsistent nurturing caused their son to suffer such a fate?

About twenty-five years ago, the pendulum in psychiatry swung fully back to the nature side. While scholars continued as always to remind us "it's always some of both," my own clinical training during this period taught me that I should reassure my patient's parents in no uncertain terms that their son's illness has nothing at all to do with his upbringing. It is caused by a biochemical imbalance completely determined by genetics. I should elicit a careful family history to hunt for even distant relatives who had hints of some similar condition, to verify both the diagnosis and its genetic basis. Thanks to outstanding teaching by some master clinicians, I learned subtle and persuasive ways to convince these parents not to have any doubts about my reassurances just because the biochemical imbalance in question has not yet been discovered.

What my teachers neglected, of course, is that in telling the parents "it's all in the genes," I am really saying "it's all from *your* genes that *you* passed on to him," and so giving them a different, but no less powerful, reason to feel guilty. Guilt is a most resilient emotion; it withstands simple attacks from doctors and other well-meaning types about as easily as a defense attorney turns the "cruel-nurturing defense" into the "defective-nature defense" in the courtroom saga imagined above.

While the position of the nature-nurture pendulum is set for each historical period by a multiplicity of factors—political, religious, scientific, economic, and so forth—the recent swing back toward the nature side resulted largely from advances in science. Although science is only one of many factors that can alter the balance, new breakthroughs in our understanding of nature itself have generally been reliable sources of support for advocates of the nature side. When scientists proclaim that they are unlocking the secrets of nature, an infectious optimism soon spreads to others who study the human condition in very different ways. Newton informs us that the universe obeys certain fixed, mathematical laws, and a century later philosophers such as Immanuel Kant are still designing nature-side arguments about how these fixed, mathematical laws govern human thought. Darwin informs us that our uniquely human capacities evolved through competition and natural selection, and immediately social theorists such as Herbert Spencer are designing nature-side arguments about how governments should stop trying to relieve the condition of the poor, since it is nature's way to let them die off if they are less fit. This influence of science has played itself out many times before, as we will see in the chapters that follow, and it is interesting to consider our current position in that light.

While advances in science generally tend to bolster the nature side of the debate, recent advances in genetics—the field claiming to hold

the secrets of our "natural endowments"—have inevitably permeated contemporary discussions of the issue. From the unlocking of
the structure of DNA and its universal genetic code to each new elegant discovery in the molecular biology of gene regulation, genetics
as a science has had a powerful influence on contemporary nature-
nurture attitudes. A single gene is discovered for Huntington's disease, and the general public soon expects that a gene (or multiple
genes) will be found to cause all other forms of mental deterioration
(including whatever is going on with that unfortunate teenage boy
in my office).

The ascendancy of genetics as a science has captured not only the
public's imagination but also its money—in the flow of federal research dollars. One of the impending triumphs of genetics is, of
course, the massive collaborative plan currently under way to sequence the human genome. I suppose I am a fan of this historic
Human Genome Project for the same romantic reasons I am a fan
of ocean and space exploration. With the technology now available
to make it possible, it is hard to see how we can resist this mapping
of our genetic endowment, any more than the early settlers of North
America could resist mapping the new continent in the U. S. Geo-
detic Survey of the last century (despite claims by some congressmen
at the time that certain large areas were not interesting enough to
warrant the expense of a massive survey). My bet is that the cost of
mapping even the most "uninteresting" stretches of our DNA will,
in the long run, reveal the molecular equivalent of the wonderful
national parks and fields of precious ore that were discovered in that
earlier survey. The doctors now in training will, in forty or fifty years,
almost certainly regale young physicians with stories of how they
trained "before we had even sequenced the genome," in the way that
wise old clinicians today talk about their treatment of patients "before we even had penicillin."

It is no accident that the Human Genome Project combines the
public's imagination with the flow of research dollars; the leading
molecular biologists who need the cash for their massive collabora-
tion are fond both of promising as yet unforeseeable cures for diseases and of making pronouncements that the sequence will teach

us "how life works" and change "our philosophical understanding of ourselves." But it is important to remember what this particular triumph of genetics will really mean. Since the genetic code is written in the four letters A, C, G, T (for the nucleotides adenine, cytosine, guanine, and thymine that constitute the DNA in our forty-six chromosomes), one way of looking at the Human Genome Project is that its net result will be a sequence of roughly three billion A's, C's, G's, and T's. Since the sequence of any person's DNA differs from any other person's by about 3,000,000 nucleotides (roughly one tenth of one percent), this product will represent a sort of "average" human genome that actually corresponds to no one real person.

Of course, scientists understand that the DNA sequence will not spell out all the mysteries of life. They know very well that any individual organism—human beings included—is a product of its unique developmental history, shaped by both *internal* and *external* forces. As molecular geneticists begin to understand how these forces interact, they have dismissed once and for all the oversimplified view that "internal" forces mean exclusively genetic factors (nature) and "external" forces mean exclusively environmental factors (nurture). After all, some of what is "external" was created by the organism, and some of what is "internal" was not part of some set genetic plan but an internal response to some external environmental conditions. What caused the genes in the beta cells in my pancreas to turn on and start producing insulin this afternoon was not some predetermined hereditary clock but my eating a chocolate bar. The fact that identical twins are born with different fingerprints despite having the same genes *and* the same uterine environment is enough to convince almost anyone of the complex interrelationships at work. Thus, while a detailed look at the DNA molecule is no doubt the best starting point for the burgeoning science of genetics, it must be said that our looking at the DNA molecule to understand "the meaning of life" is a bit like the drunk's looking for his keys under the streetlight just because that's where he can see best.

As in all other areas of the nature-nurture debate, ethical issues have loomed large in discussions of the Human Genome Project (including, in this case, the ethics of molecular biologists who promote the project for the public good while being founders, directors, and stockholders in the commercial biotechnology firms that manufacture all the supplies and equipment used in sequencing research). While capturing the public's imagination with visions of cures for disease and new philosophical insights, the specter of gene manipulation also makes people think with foreboding about the potential for prejudice, discrimination, and eugenics. Practical questions have already arisen regarding the ethics of genetic screening, since screening tests constitute some of the first promised "benefits" of sequencing research.

One such question is whether employers will be permitted to insist on genetic screening of job applicants before they are hired. The interesting arguments in favor of such testing are sometimes based upon the employer's concerns about paying higher premiums for the company's health insurance if sicker employees are hired. They are also sometimes based upon concerns for an employee who might be found to have a genetic predisposition to become sick upon exposure to something in that particular work environment. This question raises crucial issues concerning equity (what if a person now known to be genetically prone to developing diabetes is denied the job despite the best qualifications?) and also concerning industrial power in conflict with individual autonomy (what if I *want* to work at a computer screen all day even if I have a gene found to predispose me to computer-screen headache and potential lost work time?). This first set of practical questions is a mere taste of the moral issues that will be raised by human genome research in a world whose history has been punctuated by the not infrequent attempts of one tribe to exterminate another on the grounds of biological (genetic) superiority.

On the other side of the coin, we have already witnessed some of the exciting achievements in human biology and medicine that can result from advances in genetics. The prevention of thalassemia in Sardinia is a good example of a triumph of genetic screening. This disastrous genetic blood disorder is very common there, with 13 percent of the population carrying the gene. A baby has to get the gene from both parents to end up with the disease, and in 1977 thalassemia afflicted 1 out of every 250 new babies. With this high rate of disease, most people in Sardinia knew someone with an afflicted child who either died of the disorder or had to live with expensive treatment involving multiple blood transfusions and medications to control iron deposits throughout the body. A blood screening test was made available in 1977, and by 1983 the thalassemia rate had fallen to 1 in 557 births, thanks to the widespread use of this test coupled with genetic counseling and selective abortion. In 1983 a molecular test to identify the gene itself made screening even quicker and easier, and by 1989 the thalassemia rate had fallen to about 1 in 1,000 births—a fourfold decrease in this devastating disease in just a dozen years.

While this story is a triumph of genetic screening, it can also remind us how human behavior interacts with scientific discovery. In the years since 1989, an interesting trend has been developing. Since many more Sardinians now reach young adulthood without knowing someone who suffered with thalassemia, the rate at which people are getting tested has been falling precipitously. Some people predict that the rate of babies born with thalassemia will actually begin to increase again if there is no way to change the behavior of people whose knowledge of the risks involved is somehow only intellectual and not emotional. Like the human genome, this behavior is also just part of "human nature." (Another human side of the Sardinia story was the support of the local Catholic church for the screening program—an uncharacteristic position taken by priests who had also seen a lot of suffering firsthand.)

The emotional side of a genetic screening program in Sardinia does not begin to reveal the strong reaction people have to genetic manipulation—even genetic manipulation in species other than humans. The tomatoes genetically altered to stay fresh longer or the bacteria genetically altered to eat raw petroleum and clean up oil spills seem to be clear advances for everyone, and they are unquestionably only hints of exciting things to come. But even these small victories have been criticized by some with the interesting argument: "It just isn't *natural!*"

At times such as this, when the nature side of the debate is the party in power, it is hard to miss the striking way in which the "natural" is construed to mean the "proper" or the "moral." With a recent claim that scientists have found a "gay gene," the editorial pages were filled with arguments about how such a discovery proves that homosexuality is "natural"—and this is immediately taken as a decisive blow against the position that homosexuality is "improper" or "immoral." This line of reasoning itself has a long history, as witnessed by the attempts of generations of racists to demonstrate how their moral position is simply a dispassionate conclusion drawn from recognizing the inferior endowments provided by nature to the objects of their hatred. The only difference today is that we know a little bit about the genetic basis of some of our endowments, and so racists allude to inferior DNA in making this old argument. (What counts as "natural" is getting played out, ironically, in the application to genome research of the U. S. patent laws, which prohibit the patenting of anything that is "natural"—our human genome being quintessentially so, but perhaps not each isolated gene a scientist clones?)

We must not forget, of course, that during periods in history when the *nurture* side dominates the debate, the *natural* in turn becomes an object of scorn and moral debasement, as when the Reformation (and its lingering doctrines) taught people to fight their "natural instincts" and "rise above them." The notion of instincts inevitably gets tied up in this moral language. The "natural instinct" to seek sexual gratification and other worldly pleasures had a negative connotation in what became Puritan doctrine. Today, by contrast, with the nature side in domination, "good instincts" (sometimes called intuition in humans) appear to be valued almost beyond good training: just listen to the way people talk about who should teach their children or run their country!

An example which highlights how our current nature-nurture ethos has permeated thinking about the human mind itself is Steven Pinker's recent book, *The Language Instinct: How the Mind Creates Language* (1994). As a researcher who studies the development of language in children, Pinker sheds considerable light on fundamental questions concerning the constraints imposed on language by nature. Pinker's view that grammar genes determine the universal shape of all human languages becomes all the more important when we remember that modern linguistic philosophers have taken "the shape of human language" to mean "the shape of any possible human thought." I shall explain later on why I do not share contemporary philosophy's obsession with language as what ultimately sets limits on the possibilities of human experience. But it is hard not to get caught up in the questions suggested by linguistic research, especially when this research claims—like molecular genetics does—to be a direct exploration of what it means to be human.

Pinker argues for the existence of a language instinct, and it is easy to see why the issue seems at least as urgent for our understanding of human nature as the sequencing of our composite genome. If something like a universal grammar really is programmed into our brains by our genes, as Pinker argues, then this language instinct would set some constraints on how we experience both ourselves and the world around us. Here the nature versus nurture debate cuts to the heart of philosophy in its broadest sense. Despite the desire many of us ("instinctively"?) have to believe that the human mind is almost infinitely plastic and variable, along comes a psycholinguistic theory that stakes out a claim for nature's setting limits on the thoughts of which humans are capable.

This nature-nurture controversy in psycholinguistics started when Benjamin Lee Whorf, a student of Native American and other languages, made the observation in the first half of this century that people who speak Hopi have a radically different experience of time because the deep grammar of their language has no tenses or any other ways to convey the notion of enduring or lasting. Following in the tradition of the noted Yale linguist Edward Sapir, Whorf concluded that the different grammatical structures of different languages condition the way speakers of these languages think. This romantic notion, Pinker asserts, must now be debunked as a "hoax" by the language-instinct theory, which insists that our shared human genome produces one deep grammar common to all human beings, and this universal grammar determines, among other things, that people in all cultures experience time in the same way.

A genetically programmed language instinct constraining possible human experience immediately conjures up the image of biological determinism, and all the philosophical and political baggage that goes with it. As with uncovering the structure of the human genome, this image strikes fear into the hearts of people who remember how such biological determinism has been used to justify class hierarchies, colonialism, racism, and eugenics. Then again, we should remember that nurture arguments about the possibilities of human experience are just as easily put to use in power politics. Chairman Mao believed in the limitless potential of the human mind to be shaped by experience, and this belief led him to engineer China's cultural revolution. This cataclysm directly involved over 100 million Chinese people in a massive program of "reeducation" through relocation, public humiliation in "struggle meetings," torture, beatings, and very often death. As Pinker notes, the history of nurture-dominated periods reveals how a belief in the mind as a blank slate can be a dictator's fantasy as easily as a formula for liberalism.

But what, I repeat, could be more *intriguing* than questions of whether our genes shape our language and whether our language shapes our experience—especially when this experience includes the human capacities to be amused, passionate, awed, or bored and to create works of beauty, deny reality, love another person, and even lie to ourselves! The particular language I use is obviously a matter of nurture and not nature, which is why this book was written in English rather than French or German. But would its contents be more romantic if I had grown up in France, or better organized if I had grown up in Germany? And even more important, is there a common deeper structure to all human languages, as originally claimed by Noam Chomsky and now by Pinker, that must be shaping the very structure of the system of thought conveyed by the words in the book you are now reading (and any other words you will ever speak or read)?

Recently, the nature-nurture stories being told by geneticists and linguists seem to be converging, as the geneticists have discovered some correlation between people with certain genetic traits and those who speak certain languages. Does this mean that the Human

Genome Project might even solve the problems of psycholinguistics? No, as Pinker explains:

> As far as the language instinct is concerned, the correlation between genes and languages is a coincidence. People store genes in their gonads and pass them to their children through their genitals; they store grammar in their brains and pass them to their children through their mouths. Gonads and brains are attached to each other in bodies, so when bodies move, genes and grammars move together. That is the only reason that geneticists find any correlation between the two. We know that the connection is easily severed, thanks to the genetic experiments called immigration and conquest.

Pregnant with truth like so many statements of the obvious, the comment that "gonads and brains are attached to each other in bodies" really must be the starting point for any dissection of the relative contributions of nature and nurture to human thought and experience. In linking the genes found in gonads and the language found in brains, Pinker ties genetics to the other burgeoning science that (combined with recent advances in philosophy, psychiatry, evolutionary biology, and artificial intelligence) makes the nature-nurture issue so ripe for new insights today. This other new science is *neuroscience*—the study of the structure and functions of the brain. Neuroscience is itself a multidisciplinary field, combining neurobiology, neurochemistry, neuropharmacology, and just about every other scientific field that can also take "neuro" as a prefix. This new science of the brain has witnessed an explosion in the last thirty years paralleled only by the growth of molecular genetics.

While a small piece of the nature-nurture debate has centered on direct discussions of what happens to bits of DNA, most of the debate surrounds questions such as those we have already seen—the relative determinants of my daughter's jocular temperament, a criminal's immoral behavior, a mental patient's psychotic delusions, the deep structure of human thought itself. These questions come from fields as diverse as psychology, ethics, and philosophy. If ever an advance in science were designed optimally to influence this entire nature-nurture debate, the birth and development of neuroscience is surely it, with direct research implications for cognition, emotion, and motivation, which are so central to all facets of this ancient debate. We are now at an interesting point in the history of science, with the fields of molecular genetics and neuroscience *both* describing themselves as "revolutions"—a popular claim in science even before Thomas S. Kuhn codified the metaphor. (On the genetics side, the Human Genome Project has been compared not only to the French revolution but to the Holy Grail!)

I am personally captivated by brain science and shall have much to say about recent advances in neurobiology as this book unfolds, especially about the sometimes counterintuitive significance of these advances for the nature-nurture debate, conceived in its broadest context. But the argument of this book is not based in neurobiology. It is based at the *intersection* of neurobiology, psychology, philosophy, linguistics, and anthropology, and even includes aspects of sociology, history, economics, and politics. There are two reasons for inviting you—at this particular moment in time—to this exciting intersection point in order to consider issues of nature and nurture. The first is that it obeys Sutton's Law (Willie Sutton being the bank robber who, when asked by the judge why he kept robbing banks, responded: "Because that's where the money is!"). The nature-nurture debate cuts across all of the human sciences, and so their intersection is, to put it simply, where the action is.

But there is another reason for burdening ourselves with the challenges of an interdisciplinary argument, and it grows out of the grand history of the nature-nurture debate itself. During times when advocates of the nature side dominate the argument, we find in the writings of philosophers, scientists, theologians, and politicians an emphasis on good breeding, choosing good leaders, and clearing the mind of illusions that may obscure a clear grasp of the truth and the right. When the nurture side of the debate swings back into ascendancy, we find the thinkers of the day turning their focus to proper childrearing, moral education, and the mind's plasticity in reflecting the state of the world around us—for better or worse.

Given that the debate is conducted in parallel in many different—

and sometimes even competing—fields, it cannot be mere coincidence that the pendulum has tended to swing back and forth for the debate as a unified whole, even though dressed in varied garbs at each phase. There is no *obvious* reason why theologians, philosophers, political theorists, biologists, physicists, and moralists should all roughly concur at any given time on the relative contributions of hereditary versus environmental influences. Yet, the debate itself offers a frame through which we can see how various historical periods reflect an overall worldview, mirrored in the otherwise inexplicable unity found in such diverse writings on diverse subjects.

My intention in writing an interdisciplinary book is therefore not so much to review this complex and fascinating history as to move the debate forward into the future by using its history as a tool to dissect the deeper links which unite these diverse approaches to human thought and experience. In the course of this adventure, we will meet some very wise people and some very curious characters, from ancient sages like Plato and Aristotle to brilliant pioneers like René Descartes, Immanuel Kant, David Hume, Sigmund Freud, and Jean Piaget, to contemporary scientists like David Hubel, Torsten Wiesel, Stephen Hawking, Gerald Edelman, and Jonas Salk. We will even make the acquaintance of a simulated robot named for Charles Darwin.

This result will, I hope, be more than worth the journey, as our synthetic integration of insights from some of the greatest minds in history breaks down many of the distinctions most firmly held throughout that same history. Distinctions between objectivity and subjectivity, between facts and values, even between nature and nurture themselves will all dissolve, along with some questions historically thought to be vitally important, like whether science is supported by logic or logic is supported by science! I suspect that the evaporation of some of these arcane distinctions and meaningless questions will sit quite comfortably in our current nature-nurture culture, and may even, for some, come as a welcome relief.

The moral theory that emerges, in contrast, runs directly counter to the culture described above, wherein the "natural" is taken to mean the "proper" or the "moral." When we look closely, we shall discover how the moral is actually quite often the *un*natural. Is keeping your promises natural? Is *justice* natural? We put so much effort into nurturing values like promise-keeping and justice in our children. Despite current trends, I hope to demonstrate in this book why it is time to replace old-style "natural law" theories of ethics with what we might call a "nurtural law" theory of ethics. Through this new moral theory, we shall see how the triumph of molecular genetics in the Human Genome Project should in fact be a rallying point for the *nurture* side of the great debate, even as the rapid progress of that project heightens the urgency for this countercultural view of ethics.

The urgency behind our search for a new moral theory arises for two important reasons. One reason is that the Schweitzerian prospect of human gene therapy—first achieved in 1990 when a small girl was cured of a rare but lethal immune system defect when genes for a missing enzyme were spliced into the white blood cells of her body—carries with it the Orwellian prospect of germ-cell manipulations, where genes would be spliced into sperm or egg cells and so have consequences not just for one individual but for *all future generations.* This is considered the ultimate taboo among "genethicists." But no one knows yet whether genes intended for cells elsewhere in the body might ever find their way into the reproductive organs and transform germ cells. This bit of history is only now being written.

Our new knowledge of human biology is being acquired at a much faster rate than that to which the human race is accustomed. For some, genetic engineering conjures up hopeful images of new cures; for others, it suggests terrifying images of eugenics. But if by "engineering" we mean the intended replacement of some A's, C's, G's, or T's by other nucleotides of our own choosing, then only genes that have been sequenced can be engineered. As this book was being written, only about 1 percent of the human genome had been sequenced, and so not much sophisticated engineering was even possible. In just ten years, the other 99 percent of the genome will also have been sequenced, since that phase of the Human Genome Project will be complete. Although scientists will need to understand the *functions* of genes (the next phase of the project) in order to know which nucleotides they might want to replace, one short decade from now nothing will be technologically beyond engineering.

The other equally important source of urgency for a new moral theory is that, as we have already begun to see, the nature-nurture debate is much more than an academic dialogue or a challenge to biologists to uncover nature's "code of codes." In its multiple incarnations in science, politics, education, and the like, the debate has far-reaching practical consequences for social policy, law, ethics, and potentially even evolution itself. We can look around and daily see how these practical consequences affect the lives of psychiatric patients and other stigmatized groups, not to mention all school children, and, indeed, the shape of influential political and religious institutions whose guiding principles were established during some period when one side or the other dominated the debate.

My ultimate hope in articulating this synthetic view of ourselves and our world is that we can break down not only the time-honored distinctions mentioned above (objectivity-subjectivity, fact-value, and so on) but also the distinctions between the now separate disciplines of philosophy, psychology, neuroscience, ethics, and so forth. Our *tool* for chiseling away at all these distinctions will be the nature-nurture debate. Through the integrating unity of this debate, we will see in detail how psychological theories about my daughter's personality formation, ethical theories about that convict's moral development, scientific theories about the causes of that teenager's mental illness, and philosophical theories about the potential limits of human knowledge are themselves merely complementary aspects of one larger nature-nurture theory about the mutual contributions that we humans and our world make to one another. Once we have found this place from which the differences between these disciplines, approaches, and theories can be seen as merely differing perspectives—complementary points of view rather than competing points of fact—we will have come a long way toward simplifying what now appear to be complicated and even unsolvable problems concerning human knowledge and moral values.

2

The Divide-and-Conquer Strategy

The psychiatrists described in Chapter 1 both do a pretty bad job of living by what I called the *obvious* nature-nurture solution: "It's always some of both." My portrayal of an older psychiatrist acting as if "it's all in the nurturing" and similarly the portrayal of my own training to say "it's all in the genes" sounds like a caricature, created to demonstrate the extreme positions of the debate rather than the insights of thoughtful, well-educated people. Yet, we did hear doctors talk that way about nurture in the recent past and we do hear doctors talk that way about nature today. The pressures to oversimplify are staggering. They create one of the central problems to be overcome: the either-or quality that has been such a major and unproductive feature of the nature-nurture debate.

Although I have been making reference to a "nature-nurture debate," the issue is more often discussed as a question of "nature versus nurture." When one begins with nature *versus* nurture, there should be no surprise that little new light is shed on all the important issues intertwined with the matter. Only an investigation of the complex interactions between nature *and* nurture will enable us to sort out the very real (and, yes, sometimes distinct) influences of each. Hence the subtitle of this book.

Of course, many great thinkers have considered these platitudes to be plain common sense just as you and I do now: a focus on the interaction between the two is not a new approach. While it is still noteworthy that, the vast majority of the time, the debate appears to be carried on as an either-or question, let us begin by looking at some of the famous attempts to integrate the two. These attempts can all be connected by one common feature: they try to identify within us two or more separate and distinct parts, which together can account for our human capacities (to know, to love, to value, to exercise power over others) and which individually can account for the mechanisms by which nature and nurture each influence our lives.

Although some very ancient religious and mystical traditions developed something like a map of the soul or mind, the first individual to attempt a systematic account of the structure of the human mind was Plato. Plato was born into an aristocratic Athenian family in about 427 B.C. and had the good fortune at the age of twenty to become a disciple of Socrates. He enjoyed the companionship of Socrates for about the next eight years of his life, until his teacher was martyred by the citizens of Athens. Plato spent most of the rest of his life developing and teaching his philosophy at the Academy which gave academics its name, and he died at a wedding feast at his home at about the age of eighty.

Since Plato was the first to consider dividing the mind into separate constituent parts, it did not occur to him to spell out for us whether his "parts of the mind" were meant to be taken literally or metaphorically. (The controversy over this question has forced all subsequent mind-model builders to be more explicit, as we shall have to be in constructing ours.) Either way, though, his model is perhaps our best starting point. Not only was it historically the first detailed model of the mind, it was also elaborated by Plato to solve— if indirectly—the very problem of nature and nurture.

It has been said that Plato's entire philosophy is an attempt to answer the question he puts into the mouth of Meno at the beginning of the dialogue named after him: "Can you tell me, Socrates, whether goodness (virtue, excellence) is a thing that is taught; or is it neither taught nor learned by practice, but comes to men by nature, or in some other way?" Here, in words almost 2,500 years old, is the question we still ask today. And for much the same reasons. Plato lived in a time of strife not unlike our own: strife between city-states and also among the many politically divided factions within each. In seeking an answer to the problem of how people might "dwell together in unity," Plato naturally turned his attention to better forms of government and better systems of education. But before even the best state or the best teachers can provide the nurturing environment that will solve the problems of society, we need an answer to the question of whether the causes of strife are innate and unconquerable or whether people can indeed be taught to live together in harmony. It was obvious to Plato that the intellect could be educated, but could the will or spirit be taught to turn away from selfish motives and base values?

By articulating the question in more or less this way, Plato realized that he might assign different contributions of nature and nurture to different aspects of the mind, such as the intellect, the will, the spirit. With this realization came the birth of the divide-and-conquer strategy—a strategy taken up in different forms by thinkers from Plato's most famous student, Aristotle, to eighteenth-century empiricists and rationalists alike (Hume and Kant both try to divide and conquer), right up to Sigmund Freud at the beginning of this century.

Indeed, Freud's model of the mind has some similarities with Plato's, and might even be used to help understand the more ancient version of this story. Freud's familiar tripartite structural model divided the mind into id, ego, and superego, terms that have crept into our everyday language and thinking about human experience and motivation. The id is the nature side in raw form: our instincts, our appetites, and above all our sexual desires. Freud takes these as givens, as biological facts unalterable in themselves by any parental nurturing. At the other extreme he placed the superego: our conscience, the ideals we hold for ourselves that grow out of our identification with and ultimate separation from our parents. Here is the nurture side in raw form. To the job of mediating between the aggressive, asocialized impulses of the id and the socialized, moral strivings of the superego, Freud assigned the ego, whose synthetic integration of

id   ego   superego

Freud's model of the mind

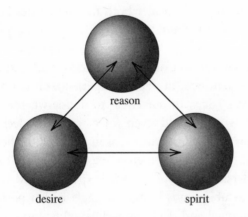

reason

desire   spirit

Plato's model of the mind

these other competing constituents makes it most identified with "me," my "self," who "I am." (Freud's original name for the ego was simply the nominative form of the German word for "I"—*das Ich.*)

Plato similarly divided the mind into three parts, which do not always see eye to eye. These he called desire, reason, and spirit. Desire is again the primary executor of the nature side. In Plato's model, when we are thirsty, desire is the part of us that bids us drink. Reason, in contrast, is the intellect, the quintessential place through which nurture acts upon the mind. This is the highest of the three and may exert its good influence over the others, as when thirsty people decide not to drink if the water is impure, because "there is in their soul that which bids them drink [desire], and also something else which forbids them, and prevails over the other [reason]." In addition to desire and reason, Plato identified spirit as a third part of the mind. Spirit is essentially our will—the seat of determination and anger, but capable of being put to good use under the direction of a well-instructed reason. Spirit can thus become a natural ally of reason against desire, as is the case in those self-controlled, stoical people who always astound more hedonistic folk like myself with their capacity to act "reasonably" at times when rage, jealousy, or some other passion would carry the day in an instant for someone like me.

It is interesting to consider both the similarities and differences between Plato's and Freud's strategies. Both view the mind as divided into distinct parts—parts which compete with one another for dominance and control over our ultimate actions and experiences. This view is consistent with our actual experience of internal conflict, as when we "struggle with ourselves" to decide whether or not to do what we know we should, even when we do not feel like doing it. Whether this struggle is Plato's reason enlisting the help of spirit to subdue desire, or Freud's superego engaging the ego to suppress the id, the model is true to our everyday experience. (My favorite metaphor for this ubiquitous human struggle is that portrayed in cartoons as an angel and devil on each shoulder, arguing over which course of action should be taken.)

While the models of Plato and Freud are similar in suggesting that both nature and nurture can influence our lives, the differences in the respective mechanisms by which this occurs are significant. Although Freud is occasionally inconsistent on this point, he generally appears to take the nature side (id) as a given. In responding to the ideals of the superego, the ego is capable of "defending" the self against the animal instincts of the id. These defenses can repress these base instincts (unconsciously), suppress them (consciously), sublimate them into more socially acceptable behaviors (creativity, humor, altruism), and employ other such strategies, according to Freud. But the id itself does not change as a result. Nature has, through evolution, given us an aggressive, sexual side; as social, intellectual creatures we can only deal with it, not change it.

Plato, in contrast, while still locating our natural, animalistic side in desire, is more optimistic about the potential for nurture to change (that is, improve) desire itself. He believed that proper schooling of the lower parts (desire and spirit) could train them to obey reason, and so result in the good ordering of our lives, which is called virtue. For Plato, this is precisely how the good life is to be attained: just as good government for all people depends upon the lower orders obeying wise rulers, the good life for an individual depends upon the lower faculties obeying a well-trained reason. Indeed, here we remember that political and moral questions really drove Plato's philosophy, and his extensive theory of education was meant not only to nurture individual minds but to train and ultimately select the philosopher-kings who would rule with absolute authority.

Throughout the ages, we can find many variations on this strategy of dividing humans into an "animal" side and an "intellectual" side in order to account respectively for the influences of nature and nurture. The variations themselves typically reveal the ulterior motives of their creators. In the two examples so far, Plato was a philosopher, Freud a physician. Plato considered the mind—with all three of its constituent parts—completely separate from the body. In writing about the educability of desire, Plato is not writing so much about training one's throat never to feel thirsty as he is about the potential for character development in order to achieve better government and social harmony. Freud, in contrast, considered the id very much connected to the body, as one would expect from a neurologist. He was driven by an interest in certain symptoms which interfere with people's lives but which appear to be generated by the very same mental mechanisms that make more successful functioning possible.

The first variation on Plato's original divide-and-conquer strategy was, of course, that of his student, Aristotle. Aristotle was born in 384 B. C., again to an aristocratic family (his father was the friend and physician of the king of Macedonia). It was his good fortune at the age of twenty-one to become a disciple of Plato. He enjoyed this companionship for about the next sixteen years of his life, until Plato died at that wedding feast. The depth and breadth of Aristotle's work over his sixty-two-year life has probably not been surpassed in these

two succeeding millennia, and his influence on history is everywhere—both through the voluminous scientific, philosophical, political, and literary writings he left us and through the considerable
impact of his teaching during his lifetime. (The three years he spent
tutoring Alexander the Great was one of the reasons Alexander was
so great!)

Aristotle ultimately rejected Plato's model of the mind because he
could not believe that human reason and desire interact in the mutual way Plato described. In its place, Aristotle developed a new
divide-and-conquer model that completely separated all desires and
indeed all motivations from the "higher" intellectual functions of
reason. By considering more carefully how these separate mental faculties could actually *combine* to produce action, Aristotle concluded
that the intellect, on its own, could never lead to *any* action. It could
discover differences between states of affairs, infer possible outcomes
of different actions, and so forth, but only some desire or preference
or some other motivational factor could account for the thought that
one such state of affairs or outcome is better or worse than another.
He therefore concluded that reason "by itself moves nothing; it is
only when it is in pursuit of an end, and is concerned with action,
that it moves anything."

Put another way, Aristotle agreed with his teacher that desires
could also be influenced by nurture: he considered it the mark of a
well-trained mind that one has learned to prefer things that are good
and actually take pleasure in leading the virtuous life. What he disagreed with was Plato's idea that reason (the best agent for nurture's
influence) could "lead" the lower faculties in any meaningful way.
The intellect can discern all sorts of differences between X and Y, but
action only results from some part of me *caring* whether X or Y
should happen. Taken to its extreme, this is the Scottish philosopher
David Hume's famous eighteenth-century inversion of Plato's roles
of reason and the passions, wherein reason "is perfectly inert" and
furthermore "is and ought only to be the slave of the passions, and
can never pretend to any other office than to serve and obey them."

A great deal has been written on this question of whether reason
is "perfectly inert" or whether there exists, as Hume's detractor

Immanuel Kant called it, a "practical reason." This crucial question has to be addressed in any nature-nurture theory that hopes to account for human experience and behavior, especially in matters of ethics. To see why so much rides on this question, we need only look back at the source of Aristotle's rejection of his teacher's particular divide-and-conquer strategy. Plato had also realized that reason on its own cannot be a motivating force in the usual sense. Occasionally he fudged the issue by making spirit the ally of reason, and so a well-trained spirit can get things moving in the right direction. But Plato's solution was generally far more radical. He actually did give reason a motivating role—by proposing the existence of an ultimate Good, the mere knowledge of which excites desire for it. It is not, in other words, anything that unusual about Plato's conception of reason itself that bestows upon it motivational power. What is unusual, rather, is his cosmology—a worldview dominated by the existence of a single, eternal, unchanging Good which is automatically an object of desire once it becomes an object of knowledge.

Whether any modern worldview can have room in it for an entity of the sort Plato's Good would have to be is a question we shall consider in later chapters. (It cannot be rejected as quickly as most moderns think.) But we can already learn two important lessons from the very existence of this subplot in the story of the nature-nurture debate. The first is about the nature of the answers we can expect. The second is about the nature of the questions themselves.

Divide-and-conquer strategies may answer some of our questions about the roles of nature and nurture, while also reflecting our actual experience of internal conflict. But they also create some new problems. Even if the issue of reason's inertness or power can be solved, we run the risk in any division of the mind of losing that single "I" that has both conflicting motives. Common sense acknowledges moments when I "struggle with myself" or "am of two minds" on a particular question. More often, though, common sense embodies more unity than disunity, and we remain skeptical of the sinner who claims he is a saint "but for the sin that dwelleth within me."

Even more important is the obvious fact that you only get out of a strategy what you put into it. All of the divide-and-conquer

approaches mentioned so far (and virtually all of them not mentioned as well) section off a piece of the mind through which the
forces of nature exert their primary influence. Whether it is Plato's
desire, Hume's passions, or Freud's id, nature appears in these theories as some kind of inherited, animalistic endowment. It is sometimes there to be educated, sometimes to be subdued, or even sometimes to inform our intellect as to which ends should be sought
through the application of our powers of reason. But it is always a
given. Once the division is made, the only question left is how we
should nurture the parts that may be nurtured. Once we divide to
conquer, there is little more to be said, except by way of description,
about the nature side.

We can therefore see why this strategy has always left nature in a
much stronger position. In those internal struggles so well described
by a division of the mind, the parts determined by nature are taken
as given, leaving the parts influenced by nurture in various subordi-
nate roles—educating, subduing, assisting. Freud's metaphor for the
relationship of ego to id was that of a small man (the ego) riding on
the back of a large horse (the id), for the relative strengths and influences are clear. Plato's metaphor had reason, the charioteer, trying
to control two horses (spirit and desire)! At best, the ego might slow
the horse down a bit, or influence its direction. At worst, in Hume's
analysis, all that is left for reason is to help the horses get where *they*
want to go.

In realizing that his divide-and-conquer strategy inevitably left
nature in the driver's seat, Plato based his entire theory of education
on that particular (limited) extent to which desire could be instructed by reason with the help of spirit. Thus, while *his* teacher,
Socrates, believed in intellectual training from a very young age,
Plato believed that intellectual training (with the exception of some
optional, morally safe mathematics) should be withheld until the age
of thirty. Until then, all training is strictly character-building: teaching people the right opinions even though it will be years before
some (and then only a select few) will be taught why these opinions
are right. Thus, only those gifted people who are capable of actually
knowing what is right will ever do more than just *believe* it, and to

Metaphors for Mixed Feelings

Plato saw reason as a charioteer trying to control two horses (spirit and desire).

Freud saw the ego as a small man riding on the back of a large horse (the id).

Modern cartoonists appear to be more equitable in their metaphor for the relative strengths of the nature and the nurture sides of our ambivalence: the devil and angel each get one shoulder!

them will be entrusted the running of not only the entire educational process, but the entire state.

One of the most impressive lessons this story teaches us is how relentlessly the divide-and-conquer strategy leads to a nature-driven outcome. Plato's *Republic* is often considered a monument to the central role of education in society—the nurturing of the young. But the *Republic* limits this education to only the highest of the three proposed social classes (corresponding to the three parts into which the mind has been divided), and the philosopher-kings regulate what little sexual activity there will be to selected pairings, with a clear eugenic purpose. Here is nature in total domination.

We have actually already seen this nature bias of divide-and-conquer strategies in the modern debate over whether there exists a "language instinct" (see Chapter 1). An oversimplified view of the question is how to assign the relative contributions of nature and nurture to human language, the language-instinct theory claiming that nature (our genome) determines our deep universal grammar, while nurture (how our parents talk to us) determines the particular vocabulary we use. As soon as this division is made, it is the genetically determined deep structure of our language that is taken to set limits on possible human thought, while the nurture-provided vocabulary becomes more of a decoration. We have already seen how the romantic notion that people who grow up speaking Hopi have a fundamentally different experience of time has to be debunked as a hoax, since the only difference between Hopi and English is the particular sound someone utters to mean "yesterday," "today," or "tomorrow." By viewing the nature side as fixed and immutable (except on an evolutionary time scale), the divide-and-conquer language theorists end any hope people might have had for further variability of the human mind, such as a capacity to experience time in some fundamentally different way.

Just as the discussion above teaches us something unexpected about the nature bias of seemingly neutral divide-and-conquer strategies, these strategies can also teach us something even more unexpected—and even more important—about the nature of the very questions we are asking. With our first scratch of the old debate's surface, we begin to see how divide-and-conquer strategies can be used to dissect hidden links connecting those diverse questions concerning the development of my daughter's jocular temperament, that criminal's immoral behavior, and the shape of our uniquely human experience of the world. From the time of philosophy's very first model of the mind, Plato showed us how a theory of moral education could devote itself almost completely to "character building"—the development of the personality. He also showed us that we cannot even begin to consider his ethical concept of the Good without addressing philosophical questions about how we can come to have knowledge of such a concept (or, indeed, about whether such a concept could ever become an object of human knowledge were such a thing as Plato's Good to exist). The ultimate inseparability of these questions about personality, knowledge, and values lies at the heart of our interdisciplinary explorations and already begins to explain the surprisingly unified voice with which the theologians, philosophers, political theorists, biologists, physicists, and moralists of a given age tend to speak on the nature-nurture debate. Like the proverbial blind men challenged to describe the large beast, these specialists each study the legs, body, trunk, or tail that all belong to the same elephant.

3

Constructing Experience

If the many diverse fields engaged in the nature-nurture debate are indeed inseparable, where should we begin? Not an easy question, since each field demands an extensive investigation, while each will remain incomplete until all the others have been both explored and then related back to the earlier discussions. This is the strategy you can expect as the chapters of this book begin to turn back upon themselves.

While there is no perfect way to assign priorities to the various disciplines we shall consider, we can perhaps do no better than to begin with basic questions about knowledge. After all, whether we want to consider the development of personality, the best form of government, or the moral way to live our lives, these fields of inquiry and all their attendant issues must first become objects of knowledge. So, even realizing that a complete philosophy of knowledge may also depend upon our answers to these other questions, it will be with the subject of knowledge itself that we begin.

In turning to basic questions about knowledge, we confront the last divide-and-conquer strategy we shall consider in this book—a strategy that has dominated the field of philosophy for over two hundred years. And if Plato's primary interest in moral and political theory shaped his division of the mind, and Freud's primary interest in personality development and symptom formation shaped his, we can be sure that basic questions about the nature and possibility of *knowledge* shaped the work of our final divide-and-conquer strategist: Immanuel Kant.

Kant was born in Königsberg in 1724, the son of a saddler (strap-maker) of Scottish origin, and it is said that he never traveled more than twenty-five miles from his birthplace during the span of his seventy-nine years. After attending college at the local university, Kant spent nine years as a family tutor before going back for graduate school and earning his doctorate in philosophy at the age of thirty-one. He spent his entire professional life as a teacher and ultimately professor at Königsberg University. Despite his parochial approach to travel, Kant's ideas radically changed the history of philosophy throughout the world.

Although Kant's German prose is generally considered obscure at best, his theory of knowledge is not all that complicated. In order to understand Kant's model of the mind, we need to step back for a moment and look at the specific problems that he was trying to solve. At the time Kant was active, it had already been about 150 years since the French thinker René Descartes had drawn the attention of philosophers to the many ways we humans can be wrong in our beliefs about the world. In particular, Descartes highlighted the distinction between the *things* that we claim to have knowledge of (things that exist "out there in the world") and our *thoughts* about those things (thoughts that exist in our heads, and arise from some interaction between ourselves and the things "out there" we want to know). By drawing our attention to the difference between thoughts and things, and also to that crucial *interaction* that can distort our view of things (as seen most clearly in cases of optical illusions), Descartes challenged future philosophers to find any possible way to be sure of their knowledge—to be sure, in other words, that their thoughts about things reflect the real truth of those things.

By the time Kant began working on this central question about knowledge, philosophers over the century or so that followed Descartes had reframed it more precisely as a question about the various features that we attribute to things, and particularly about where these various features *come from*. We attribute many qualities to the

The Müller-Lyer Illusion

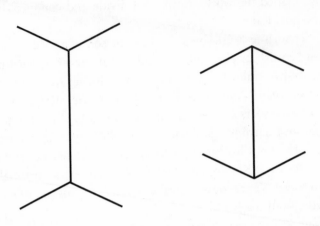

The apparently unequal length of the equal Müller-Lyer arrow shafts highlights the *interaction* between ourselves and the objects we experience in the process of coming to have knowledge of those objects.

things we claim to know, to the "objects" of our experience. We consider them to be red or blue, round or square, moving or at rest. We consider them to exist in space and time (to be spatial and temporal) and to obey certain causal laws of nature. We also consider them to be beautiful or ugly, useful or worthless, delicious or disgusting. But once we realize with Descartes that all our experiences of objects arise only from some interaction of ourselves with those objects, the question immediately arises as to which of those qualities really *come from the objects* in question and which are really *contributed by us* in the process of interacting with them as we come to know them. If these qualities were limited to shapes, colors, motions, and even perhaps temporality, spatiality, and causality, it would be easier to believe that the objects in question contribute all of the features we claim to know about them. This is the commonsense view, since it is, after all, about *them* we are claiming to have knowledge. But the introduction of features like beauty, usefulness, and tastiness makes it obvious that we contribute to at least some of the qualities we attribute to things in the world. But exactly *which* qualities come from us?

As Kant began to consider this question of how the features of our experience should be apportioned between ourselves and the objects themselves, he realized that the commonsense view (that all important features are contributed by the objects themselves) simply could not be true.

One of the strongest cases for that commonsense view had recently been made by the Scottish philosopher and historian David Hume in his *Treatise on Human Nature* (1740). David Hume's sixty-five years (1711–1776) were enjoyed entirely within the eighteenth century, a time of great optimism about the limitless capacity of the human mind. This nurture-dominated period spawned not only a lot of philosophical empiricism but also the American and French revolutions, based on such "self-evident" truths as egalitarianism, liberalism, and individualism. Hume was born in Edinburgh and was originally trained as a lawyer. He did not like practicing law, and gave it up for a rather colorful life as a scholar and socialite, traveling extensively throughout Europe and befriending most of the other intellectuals associated with the period (Diderot, Rousseau, and the rest). Although his *Treatise on Human Nature* "fell dead-born from the press" (Hume's words), it went on to be recognized later as the greatest extension of the empirical philosophy originated by John Locke, who had summarized empiricism's nurture-dominated view in this much-quoted passage from *An Essay Concerning Human Understanding* (1690):

Let us suppose the mind to be, as we say, white paper, void of all characters, without any ideas:—How comes it to be furnished? Whence comes it by that vast store which the busy and boundless fancy of man has painted on it with an almost endless variety? Whence has it all the materials of reason and knowledge? To this I answer, in one word, from *experience*. [emphasis in original]

In keeping with the nature-nurture ethos of his buoyant day, Hume followed Locke and maintained that the mind is something like a blank slate, a *tabula rasa*, on which the world writes itself.

While it was relatively simple for Hume to believe this for those features of experience directly available to perception (colors, shapes, motions), when it came to the more conceptual features of experience (temporality, spatiality, causality) Hume recognized that he had a problem.

It was clear to Hume, as it is to us, that we do not experience some features of the world as directly as we experience others. Consider the experience of seeing a rock breaking a window, for example. The color of the rock, its trajectory, and the shape of the resultant hole in the window are all readily perceptible. But consider now two other features we experience: that the rock *caused* the window to break and that the rock is an *object* that persists through space and time.

Hume realized that we never really *perceive* the "causation" we attribute to the rock's breaking of the window, and so it became difficult for him to include causation in his commonsense analysis of how all the important features we attribute to objects are contributed solely by them to our experience. Indeed, he quickly realized that causation is a concept *we bring to* experience, since we never actually *find it in* experience as we find colors, shapes, and other perceptible sensory features of objects. Specifically, Hume said that, having seen enough rocks break enough windows, our minds then add the quality of causation to the experience, perhaps out of some tendency to generalize our experiences and perhaps to reassure ourselves that the future will continue to exhibit this same relationship between rocks and windows (even though we can never really be sure).

What this means is that causation *does not really exist* in the world. For Hume, that feature of experience, since it could not be attributed to the objects in question, must be a fiction that we invent and spread over our experiences as we interact with the world. The inevitable conclusion of the nonexistence of certain important features we usually attribute to our experience was, needless to say, something of a blow to a theory claiming to support itself on the basis of common sense.

It was Kant's recognition of Hume's error on this point that Kant tells us "awakened him from his dogmatic slumbers" to appreciate the complex truth of the matter. In order to see how Kant reacted to Hume's attempt to solve this allocation-of-qualities problem, we need to make use of the well-known philosophical distinction between the "a priori" and the "a posteriori." *A priori* means "before experience" and implies that something could never be found to be true *or* false through experience. *A posteriori* means "after experience" and implies just the opposite: the a posteriori is the empirical—that which can be confirmed or falsified on the basis of experience. Thus, although Hume found certain conceptual features of experience (such as causation) to be in us (and not in the objects "out there"), he claimed that these features of experience were in us a posteriori. It was, in other words, only as a result of multiple experiences of seeing rocks break windows that our mind comes to add this quality of causation to the interaction.

In likewise appreciating that these conceptual features of experience cannot be found in the objects "out there," Kant could only agree with Hume that they are to be attributed to us and not to the objects we experience. Where Kant disagreed with Hume, however, was in the story he told about how we come to add these qualities to those objects as we experience them. Kant said that these qualities (temporality, spatiality, causality) are in us, but *in us a priori*. We do not "add" the quality of causation because of our past experiences with rocks and windows—we rather experience a rock breaking a window as the causal process it is because we apply the concept of causation to the data of our experience (the shapes, colors, motions) in the course of constructing our complex experience from that simple sensory data!

Kant's realization that we "construct" our experience (and do not merely passively "receive" it, as the empiricist Hume suggested) led to the development of history's most intriguing divide-and-conquer model of the mind. In apportioning to the objects of our experience only the crudest sensory data and apportioning to *us* all those complex intellectual qualities that make experience so interesting, Kant divided the mind into two main parts or "faculties." The faculty of *sensibility* was the part of the mind that passively receives the sensory information from the world about red or blue, round or square,

Kant's Model of the Mind

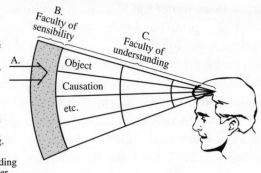

A. Objects "out there" in the world contribute sensory data, such as shapes, colors, motions, and textures.

B. These sensory data are received by the faculty of sensibility, which organizes them and presents them to the faculty of understanding.

C. The faculty of understanding throws a conceptual grid over these simple sense data and organizes them into our complex experience of the world according to concepts (categories) such as "object" and "causation."

moving or at rest. The faculty of *understanding* had the more active job of organizing those sensory data into our experience by throwing over those data a grid of concepts that the mind uses in constructing its experience.

As an example, consider the other feature of the rock-breaking-window experience that I mentioned above: that the rock is an *object* that persists through space and time. The world throws lots of sensory information at us, but why do we experience these shapes, colors, and motions as belonging to fairly permanent objects? Just as we do not perceive causal relationships directly, we do not perceive the object-ness of objects in the same direct way we perceive their sense data (shapes, colors, and so on), yet we carry on as if all of those sense data come from fairly permanent objects into which the world is carved.

A Humean response might be that because things "hang together" so consistently, we eventually start to assume (a posteriori) that this is a feature the world actually has. Kant would answer that we naturally always experience the world as made of objects because "object" is a concept we always *bring to* experience (a priori) in constructing our experience from the sense data available to us. Similarly with temporality and spatiality. It is true that every single experience we have is of a spatial, temporal world. But, Kant would say, this only proves the point: we never have experiences outside of space and time precisely because these are conceptual features that the mind always applies in constructing any possible experience of the world.

Having decided that Hume was wrong not in assigning these conceptual features of experience to us but in assigning them to us a posteriori, Kant set out in his *Critique of Pure Reason* (1781) to determine exactly which features of experience should be included in the grid that we throw a priori over the world as we construct our experience of it. Specifically, Kant set out to determine a minimally sufficient list of such a priori concepts that could account for all the features of experience we have. He adopted a term from Aristotle, "categories," to refer to these experience-shaping concepts, and deduced that object, causation, and ten others could account for the form human experience necessarily takes ("necessarily" because we must always apply these categories in constructing any experience we can have).

I write "the form human experience takes" because one way to understand Kant's divide-and-conquer model of the mind is to distinguish what might be called the *form* of experience from what might be called the *content* of experience. In essence, Kant said that in allocating the various features we find in experience between ourselves and the objects out there in the world, *we* supply the features that shape the form of experience (through the application of the categories by our faculty of understanding). The *objects themselves,* in contrast, supply the content of experience, by filling in our conceptual scaffolding with the sensory contents received from the world by our faculty of sensibility.

Where then would Kant say these form-determining categories of experience actually come from? From us—from our *nature.* Although as a philosopher Kant was not actually all that interested in the contingencies of biological reality, we might understand Kant's division of the mind into the faculties of sensibility and understanding as a division of those features of experience contributed by the environment (the sensory data offered to the faculty of sensibility by the world) and those features of experience determined by our human nature (the conceptual grid of understanding we always apply in constructing and thereby shaping the form of that experience). Simply put, Kant might say that nature provides the form, nurture the content. It is in the nature of being human to apply the categories of understanding in the process of constructing experience—which is why Kant called his categories the "necessary conditions of any possible experience." As we discovered in the previous chapter, this divide-and-conquer approach naturally leaves nurture with a considerably less interesting job: that of keeping sensibility supplied with shapes, colors, motions, textures, temperatures, and so on, so that understanding can have some content for the categories to form into our experience.

This is not the usual way of looking at things. When, say, a physicist investigates the causal connections that hold between objects, we usually would say that this scientist is attempting to *describe* the causal laws of nature. If those causal laws of nature are not contributed to our receptive faculty of the mind by the objects in the world, but are contributed by our own constructive faculty of understanding as Kant said, then, to use Kant's own words, "the understanding does not derive its laws from, but prescribes them to, nature." Kant himself realized that this view is counterintuitive, adding that this "seems at first strange, but is not the less certain." Indeed, Kant would remind us that the *universality* of the laws of nature are ensured only because we apply them a priori in constructing experience. They can never be falsified in experience by physicists or anyone else because "they are not derived from experience, but experience is derived from them." Indeed, it was not just any old laws of nature that Kant believed are built into the mental grid we use to construct our experience of the world, but those very laws of nature that Isaac Newton had described in his *Principia* in 1686. Here we see Newton's scientific discoveries still supporting nature-side arguments about the mind a century after his work was published, as Kant located in the form of our *thought* the laws of physics Newton believed he had found in the *world*.

The similarity of this story to our earlier discussions of language should come as no surprise. When the notion of a deep universal grammar was first outlined by Noam Chomsky beginning in the late 1950s, he self-consciously adopted Kant's model. Our nature-determined, language-shaping "grid" organizes the *form* (deep grammar) of any language we humans can possess, while the *content* of our vocabulary gets supplied by whatever words are offered by our particular nurturing environment. As contemporary philosophers have focused on language as the critical factor shaping human thought and experience, Chomsky and his followers (including Steven Pinker) have inherited Kant's philosophical position, with its unusual divide-and-conquer model of the mind and its constructive theory of experience.

Support for Kant's "constructive" theory of experience has recently come not only from linguists but also from experimental psychologists and neurobiologists. That is, as we learn more about our mental processes and about the brain mechanisms that give rise to them, more and more evidence is mounting to support something like Kant's model, wherein sensory "inputs" are "processed" by faculties using conceptual grids to organize these inputs into a coherent experience. I shall have much to say later in this book about what this recent evidence means for our struggle to apportion various aspects of experience between nature and nurture.

For now, before going on in the next few chapters to try to turn Kant's theory on its head (in exactly the same way Kant did with Hume's), I would like to conclude this introduction to Kant's model with two final points. The first is an observation about one way in which Kant's strategy offers us much more than our previous divide-and-conquer models of the mind. The second is a comment about the role of reason in all that will follow.

Although I have characterized Kant's model of two mental faculties as a divide-and-conquer strategy, Kant considered sensibility and understanding to be as inseparable as they are distinct. In assigning the separate roles of nature and nurture not to desires or passions versus intellect or conscience but to *form* versus *content,* Kant created two mental faculties that are much more intimate than in our previous models. Indeed, using the term "intuitions" for the sensory content brought by sensibility to understanding, Kant wrote,

> *Thoughts without content are empty, intuitions without concepts are blind* ... These two powers or capacities cannot interchange their functions. The understanding can intuit nothing, the senses can think nothing. Only through their union can knowledge arise. But that is no reason for confounding the contribution of either with that of the other; rather, it is a strong reason for carefully separating and distinguishing the one from the other. [emphasis added]

I quote here directly from the *Critique of Pure Reason* because the relationship Kant describes is crucial to understanding the links between nature and nurture in any theory we would want to develop. Plato's and Freud's models separated the two so much that we can imagine the philosopher-king or completely socialized individual who has no desires or aggressive impulses, whose nature side has completely given way to a well-instructed or well-developed nurture side (reason/superego). Alternatively, in those same theories, it is quite possible to imagine nature on its own in an uninstructed Platonic slave following only his desires or in a psychopath unencumbered by the conflicts of a superego. By resting his argument on the distinction between form and content, Kant claimed that *no* experience is possible—even in principle—without the reciprocal contributions of both nature and nurture: hence his conclusion that they are as inseparable as they are "distinct" (identifiable as individual contributors to the integrated whole). This carries over in the modern linguistic adaptation of Kant's model, wherein *no* language is possible without the contributions of both an innate grammar and a learned vocabulary. With apologies to Kant, we might say that "grammar without vocabulary is mute, vocabulary without grammar is meaningless!" As we develop Kant's model in ways he would never have imagined, we shall in fact see how form and content, along with nature and nurture, are even more intimately intertwined than even Kant recognized.

Finally, we cannot leave Kant's model without at least a brief comment on the title of his famous book. If the *Republic* tipped us off to Plato's ulterior motives in a philosophy aimed at the quest of social harmony, how did a quest for "pure reason" influence Kant's model of the mind?

Kant actually divided the mind into several faculties, not just two. For example, the faculty of imagination worked with understanding to fill in gaps left by sensibility. Most important, though, was the faculty of reason, which was distinct from the faculty of understanding. Kant was clear that the faculty of understanding could never come to any knowledge or experience without applying its "categories" (concepts) to the empirical content offered by sensibility. The application of these categories was thus bound to empirical content, without which, as Kant said above, they are empty. In contrast, Kant assigned the faculty of reason the job of manipulating the categories themselves, apart from any empirical content that might tie the process to the outside world of "things." It was in this sense that he claimed that reason could be "pure"—pure in that it was not contaminated by the a posteriori world (as even his a priori categories had to become contaminated in their necessary application to the a posteriori contingencies found in the content of sensibility).

Whether reason itself can be pure in this sense—can be, that is, completely disconnected from the contingencies of the world of things "out there"—is a question of central importance to the nature-nurture debate. For Kant, the categories were offered by nature, but they only became meaningful when mixed with nurture (the intimate link described above). Reason, in contrast, is Kant's last bastion of pure nature, completely unaffected by the contributions of the "environment."

As we proceed, let us keep an eye on the purity of reason. If nature and nurture are as intimate as they might be, perhaps even reason itself will become "contaminated" by reality.

4

Thoughts and Things

While Hume saw the mind as a blank slate or *tabula rasa* upon which the world can write itself without distortion, Kant pictured the mind as more of a piece of contoured marble whose own features contribute in no small part to the picture of the world that gets etched upon it. In order to understand why Kant's model so transformed theories of knowledge ever since, it is useful to contrast his view with that of Descartes, the other philosopher besides Hume whom Kant sought to debunk. A look at Kant's attack on Descartes will also put us in a good position to move beyond Kant to Hegel, who took up the very weapon Kant invented to critique Descartes and leveled it at Kant himself.

René Descartes was born in 1596 in France in a small town near Tours. He spent most of the first half of his life in France and most of the second half in Holland, although vexatious controversies with Dutch theologians induced him to accept an invitation from Queen Christina to move to Sweden in 1648, and he died in Stockholm two years later at the age of fifty-three. When students today first come to Western philosophy, they traditionally start with the work of Descartes, who is generally regarded as the "father" of modern philosophical thought. It would be nice if this honor were given Descartes because he took the first steps toward solving those philosophical problems which occupy the modern mind. Unfortunately, he did not do this. Quite the contrary: like altogether too many parents, Descartes is remembered for getting us *into* those problems which we are still struggling our way out of.

Descartes's famous "method of doubt" is said to have occurred to him at the age of twenty-three while serving as a soldier, stationed at Neuburg on the Danube. It was about ten years later that he began his remarkable twenty-year period of great seclusion, meditation, and writing. In his philosophical work, Descartes wondered about the veracity of his beliefs, and he set himself an amazing challenge: Is there anything I can really be sure of if I continually ask the question, "But how can I be sure of *that*?" Not even allowing himself to be sure that an evil demon is not intentionally and continually trying to trick him, Descartes quickly discovered how far such skepticism can lead. One might wonder whether *anything* could withstand such a rigorous test of certainty, but Descartes did eventually discover at least one thing that could never be doubted. This one thing he summarized in what has become the most famous sentence in all of philosophy: *cogito ergo sum*—I think, therefore I am.

What Descartes realized he could be sure of was that he existed— existed, that is, at least as a skeptical doubter doubting everything. The reason for this famous conclusion is actually quite simple and flows from one special feature of thoughts: they remain thoughts even when they are doubted. (In fact, even when they are proved wrong, thoughts are still thoughts; they are just incorrect thoughts.) In doubting everything he could, Descartes discovered that he could not doubt the existence of his own doubts, since doubts about them are doubts themselves! But to prove only that one exists as a skeptical doubter is hardly cause for celebration, and most of Descartes's philosophy of knowledge is actually an elaborate theological argument that derives the truth of the existence of everything else essentially by saying: "As I am an imperfect being, perfection could not arise from me, so I could only have the clear idea of a perfect God if a perfect God put it there, and such a perfect God would not deceive me about all my beliefs about the world, so everything in the world surely exists as it appeared to me before I started doubting it all in the first place."

Descartes was a brilliant thinker whose theories about physics were superseded only by Isaac Newton and whose mathematical inventions included the equations, exponents, and analytical geometric coordinates we still use today. But as his *Meditations* (1641) deduced the certain truth of $2+3=5$ much more effectively than the truth that "things outside us" exist, it comes as little surprise that Descartes's theory of knowledge has not withstood the test of time among philosophers. It was, however, such a radical philosophical point of departure for challenging the veracity of our beliefs that most theories of knowledge since have defined themselves in Descartes's terms, even as the philosophers developing these theories try to set him up as a straw man to advance their own arguments against him. The modern critique of Descartes is perhaps best captured by the contemporary scholar Frederick Will, who said of the Cartesian philosophical tradition:

> It is hard to view a method as a promising way of dealing with difficult problems in the understanding of subatomic particles, or claims made about the supernatural, if after centuries of trial it has been unable to give a remotely plausible answer to the question whether there are dogs and cats.

In order to see why Descartes continues to have such dramatic impact on philosophy, we need to appreciate how he relied upon the beneficence of God for an even bigger job than that of guaranteeing the truth of all our beliefs about the world. One of these (presumed to be true) ideas Descartes had about the world was that *thoughts* (the mind) are "pure consciousness," while *things* (matter) are "pure extension." Since these attributes appeared to Descartes to be mutually exclusive, mind and matter (including the body) could not really affect each other without the *continuing* intervention of God. If human thoughts have any connection with, say, neurons in the brain, it is merely an occasion of God's handiwork.

Here we see the dualistic problem Descartes left for modern philosophy. Nowhere is Kant's brilliance more evident than in his approach to solving it. The problem as Kant saw it was not so much that Descartes's thoughts failed to prove the existence of "things" (like dogs and cats), but that his thoughts were so totally *independent* of things that they never even stood a chance. By focusing his entire project just on his own inner, subjective world of thoughts, it is hardly surprising that Descartes was never able to say much about the external, objective world of things. The unfortunate (secular) conclusion that "I can only be sure that I exist" is only one small byproduct of the striking independence of thoughts from things that is so much a part of the Cartesian tradition.

Unfortunately, this Cartesian tradition continues to shape much of the modern view of human experience. By focusing all concern on the individual's experience, Descartes established a methodology doomed to failure. Since his own thoughts, his own inner experiences, were ultimately the only matters of interest, the only matters to be found doubtful *or* certain, Descartes established the individual's *subjective* experience as the yardstick by which to measure the veracity of beliefs, by which to determine the *objectivity* of knowledge. Once locked into this inner subjective world, it is no wonder that philosophers following this tradition have been struggling ever since to get back out. By even entertaining the possibility that I could have the same *thoughts* I do without the external world of *things* even existing, these thoughts are taken to have such a radical independence from things that human consciousness becomes nothing more than a private thought-theater.

The Freudian model of the mind that we saw above, as a post-Cartesian theory, is a good example of how this tradition has permeated the modern mind. For all of his revolutionary insights into the human condition, even the physician Freud carried on in a philosophical tradition that accepts without question this view of the mind as a private thought-theater. True, Freud was one of the first to realize how important backstage (unconscious) events are to the "show" that appears in the theater's centerstage. But one can summarize concisely most of the criticisms now leveled against him by saying that Freud focused too much on the inner world of his patients' *thoughts*—their conflicts, wishes, fears, and fantasies—and not enough on the outer world of *things*, like perpetrators of incest, abuse, and other "external" contributors to the show that plays in that theater of the mind.

Kant sought to solve this problem by addressing himself to the source of this striking Cartesian independence of thoughts from things. The source of the problem, Kant realized, was an unusual feature of Descartes's model of the mind: its *passivity*. So long as the mind is just an empty thought-forum, passively waiting to be filled by whatever it is the world has to offer, there can at best exist only a tenuous connection between our thoughts and the things that really exist in the world. The Kantian model described in Chapter 3 attempts to solve just this problem. For Kant, the mind is not merely a thought-containing object but an active participant in that which it knows.

By making knowledge an *activity,* and not merely a *product,* of the mind, Kant sought to reunite thoughts and things. In the Kantian model, no experience of the world is possible unless the faculties of sensibility and understanding have already collaborated to organize the sensory contents of the world into the conceptual form we humans use to make sense of the world. Kant thereby hoped to show that one could not begin coherently to doubt the existence of things as Descartes thought he had, since any thoughts we have are only made possible through the mixing of our thoughts with those very things. Kant's model of the mind is built upon the active integration of the form and content of experience through the active interactions of his faculties of understanding and sensibility. Without both form (provided by us in the categories we apply to experience) and content (provided by things in the world), *no* experience is possible. Although it did not seem obvious at first that Cartesian doubts are in themselves self-refuting (as witnessed by generations of first-year philosophy students who return from their classes wondering whether the tables and chairs in their dorms "really" exist), Kant showed that any experiences of the world we can ever have are—in a nontrivial way—thoughts about things.

In introducing Kant's interactive model, I used the notion of contributions of *things* out there and contributions of *thoughts* from ourselves as separate, identifiable sources of the various features of our experience and knowledge of the world. We can now better understand the nature of these reciprocal contributions. Once knowing has become an activity and not merely a product, we can view the contributions of thought itself as whatever form-determining features have been brought to our experience by active application of our mental structures, be they understood as Kantian faculties, Freudian ids and egos, or neurobiological brain pathways. In all of these models of the mind, there are some important contributions to our ultimate experience of the world that come from the structure of thought itself. This is almost common sense, when we consider how two different people can look at the same "thing" and have two totally different experiences of it. The different contributions to these experiences from their different mental structures leads to very different "thoughts" even when the sensory contributions from the "thing" in question is presumably the same for both of them.

But what can we say about the contributions of things to the interaction? On the one hand, Kant was unquestionably correct that things "contribute" basic sensory information about themselves, so that we come to know them as red or green, round or square, moving or at rest. But, on the other hand, if we really want to say that our thoughts are thoughts about things, we would like to say that all of those other interesting qualities (object, cause, space, time) are *also* contributed by the things in question and are not merely by-products of our minds. Otherwise our thoughts are mostly thoughts about thoughts!

There is no question that we do, as Kant said, apply concepts to our experience in the course of interacting with the world. If we are ever going to be able to say that our experience of things really is experience *of things,* we are going to have to find a way for the things to contribute to the shape of these very concepts, and not merely contribute colors, shapes, and motions. The Hegel scholar Frederick Will (whose comment above about the existence of dogs and cats was intended to include Kant in the Cartesian tradition he was criticizing) summarized this point by saying that we have to find a way in which "having a concept of a thing is dependent upon things." We need to discover, in Will's words, "how we think with the help of things."

Will grounds this challenge in a tradition that begins with Hegel because Hegel was the first to level Kant's criticism of Descartes right back at Kant. Kant said that Descartes's thoughts were too passive, and as such too independent from things. Hegel in turn looked at Kant's thoughts and saw that Kant only went halfway toward solving the problem. True, the faculty of understanding is active enough— but totally shaped only by thought itself. The faculty of sensibility, in contrast, is intimate enough with the world of things, but is (like Descartes's entire model of the mind) too passive, sitting and waiting to receive whatever sense data the world of things has to offer. Since the "real" conceptual truth of these things is isolated from Kant's active, form-shaping faculty, Kant is left having to admit that the world-as-we-know-it (always a temporal, spatial, object-containing world) has little to do with the world-as-it-really-is (which may have none of these properties). Sensibility may be in close touch with things, but it is too passive to absorb most of their interesting properties. Understanding may be active, but it is shaped only by thought (by *our nature*) and is therefore as independent from the real truth about things as were all of Descartes's thoughts.

We might summarize Hegel's critique of Kant by saying that Hegel found Kant's concepts (the categories applied by the faculty of understanding) to be too fixed and rigid. Hegel agreed that we must apply concepts in the course of constructing our experience of the world, but he saw these concepts as much more fluid than did Kant. We do apply a conceptual grid in constructing our experience, but the shape of that grid can itself be changed as we continually apply it in interacting with "things." Hegel saw the mind as capable of constantly examining its own concepts in the light of its total experience of the world, so that the world of things could help us shape—and *improve*—our concepts. While with Kant knowledge became an activity and not merely a product, Hegel made knowledge an *achievement*, and an achievement that requires constant vigilance to be sure our current concepts incorporate as much of the world as they possibly can.

Hegel was a German philosopher who was born in 1770 and died of cholera at the age of sixty-one. In addition to his influential work on logic and aesthetics, Hegel developed an intriguing view of the evolution of history as a progress by antagonism in which conflicting principles inevitably become subordinated to a deeper unity they both express. Karl Marx's adaptation of this dialectical philosophy helped it become a major force in world history. On the issue of the relationship between thoughts and things, however, Hegel is usually taken to have been a philosophical "idealist," believing that the entire world of things is actually just the manifestation of thoughts. We would probably not want to begin a synthesis of philosophy, psychology, and science with the view that the material world is merely the way thought realizes itself, and Hegel's prose is unfortunately even more obscure than Kant's (which is saying something). But Hegel's realization that knowledge can improve and progress is one of the most important discoveries in the history of philosophy. With his fluid view of our most basic concepts and his holistic view of human experience, Hegel would say that we can examine our concepts in the light of that total experience. If our current concepts are not supported by contributions from the world of things, then we should be able to discover in our thoughts subtle inconsistencies or inadequacies that will help us improve them.

In order to see how this process for improving our concepts works at a practical level, I shall demonstrate it for one such concept in detail in the next chapter: the concept of object. Before elaborating on that example, however, it is worth making one comment on the role of reason, an idea introduced at the end of the preceding chapter. You will recall that Kant was very impressed with the possibility of a "pure reason." This meant that the faculty of reason had the task of manipulating a priori concepts without any reference to the external world of things. While the faculty of understanding was not "pure" (since it could only apply concepts to the contingent, a posteriori data offered to it from sensibility), Kant's faculty of reason could operate in as much isolation from the world as did Descartes's entire model of the mind.

Hegel rejected Kant's notion of pure reason, since he saw even our basic concepts as having a larger context within which they could be examined. For Hegel, even reason itself cannot exist in isolation from the world of things. That is, whether there are or are not a priori features of experience (determined by thought simply as thought), there must also be, as Will puts it, "a revelation in thought of the character of things because of the role of things in supporting, guiding, making our thoughts possible." Put another way, Kant's reason was not really "pure" from the perspective of our modern view because, in that it was concerned with the innate set of a priori concepts Kant assigned us, reason must be conditioned by the structure of thought itself. For Kant, reason may not have been contingent on the world of things, since it was, as we reframed his argument, contributed by nature, not nurture. Hegel, in contrast, balks at a reason that is relative only to the structure of consciousness, and substitutes for this the view that *reason is relative to reality.*

In practical terms, what Kant overlooked is that *we* (and our mental faculties) are, like the objects of our experience, also in the world. Our nature (including our faculty of reason) is as much a "contingent" fact of the existing world as anything, and as the world changes, so must our conception of it.

Since Hegel, philosophy itself has come to be understood as an unfolding story, developing as history develops, and not some set of "self-evident" truths. Hegel's primary metaphor is that of *growth*—in sharp contrast to the Kantian image of the mind as a steel filing cabinet of fixed concepts. Biology (with its notions of growth, development, evolution) now replaces physics (with its more static and immutable laws and equations) as the dominant philosophical image, so that knowledge develops not like a mathematical deduction but like a tree blossoming. This image appears in the first pages of Hegel's *The Phenomenology of Spirit* (1807) and dominates all of his thought:

> The bud disappears in the bursting forth of the blossom, and one might say that the former is refuted by the latter; similarly, when the fruit appears, the blossom is shown up in its turn as a false manifestation of the plant, and the fruit now emerges as the truth instead. These

forms are not just distinguished from one another, they also supplant one another as mutually incompatible. Yet at the same time their fluid nature makes them moments of an organic unity in which they not only do not conflict, but in which each is as necessary as the other; and this mutual necessity alone constitutes the life of the whole.

I encourage you to read through this passage several times, because it contains the seeds (as it were) of Hegel's entire system of thought. Hegel's biological metaphor can serve as a constant reminder that, in contrast to Kant's insistence on a pure, rational world of knowledge, ours is a world filled with wonderful contradictions. These are not logical contradictions but simply differences in form: a bud cannot be a blossom or a fruit at the same time, yet it becomes all three through successive moments in the life of the whole.

In a description of his own philosophy, Hegel wrote of his *Phenomenology*, "This volume deals with the *becoming of knowledge*," a phrase that captures his holistic belief that everything short of the whole is fragmentary and incapable of existing without contradiction unless complemented by the rest of the world. We saw above Hegel's claim that even our basic concepts can *become*: "if our current concepts are not supported by contributions from the world of things, then we should be able to discover in our thoughts subtle inconsistencies or inadequacies that will help us improve them."

The great twentieth-century philosopher Bertrand Russell developed a wonderful metaphor for this Hegelian process by which our faculty of reason can search even our basic concepts for those subtle inconsistencies and inadequacies that will help us improve them and keep them linked intimately with reality:

Just as a comparative anatomist from a single bone sees what kind of animal the whole must have been, so the metaphysician according to Hegel sees from any one piece of reality what the whole of reality must be—at least in its large outlines. Every apparently separate piece of reality has, as it were, hooks which grapple it to the next piece; the next piece in turn, has fresh hooks, and so on until the whole universe is reconstructed.

Hegel did not believe that reason could be pure—in the sense of being "uncontaminated" by the contingencies of the world—because he understood that the role of reason was precisely to search our basic concepts for such contradictions and incompleteness in order to achieve an ever more complete view of reality.

This process is well suited to our task, and we shall return to the image of metaphysical hooks again and again as we become the comparative anatomists examining successive conceptions of reality to discover what internal contradictions and inadequacies might lead us to a more complete understanding of the world and so "uncover in our thoughts the signatures of the helping hand of things," as Will put it. Ultimately, these successively broader contexts will also point us to the subtle links that unite diverse fields such as philosophy, psychology, ethics, and neurobiology into a unified whole—a reminder that our brief exploration here of the nature and nurture of our knowledge will itself remain incomplete until we have explored other fields and then returned to these questions discussed by philosophers. In an important way, then, this book is not just *about* this process—this Hegelian "becoming of knowledge"—this book exemplifies that process.

Returning, finally, to the language of nature and nurture, we might say that Kant's view was a little too tidy. With his fixed divide-and-conquer faculties of the mind, Kant believed that our basic concepts are contributed strictly by our nature, the nurturing of "things" being limited to the sensory contents of experience. If now we seek to understand how having a *concept* of a thing is dependent upon *things*, we must muddy Kant's clear waters with some degree of nurturing of even our basic concepts of the world. Let us take a closer look at one example of such a concept.

5

How Babies Think Things Over

Kant discovered the active way in which the structure of thought contributes to the shape of our experience of things, and every theory of knowledge ever since has had to take this truth as a starting point. Above, we considered this discovery in terms of Kant's rather unusual apportionment of the contributions of his two faculties, and so of nature and nurture, to the resultant experience of the world produced by the interaction of these faculties and that world. Through the nurturing of the environment, the world of things contributes the content of experience: the crude sensory data of shape, color, texture, temperature, and the like. The nature of mind itself (through the active work of its faculty of understanding) contributes the form of experience: all the conceptual features of spatiality, temporality, object, causation, and so on. If Kant is right, then the environment contributes only raw sensory information to our knowledge of the world, while our own nature contributes all those fascinating intellectual features about which we are generally more interested. (Here we see the nature bias of divide-and-conquer strategies yet again.)

As mentioned in Chapter 3, this is hardly the usual way of looking at things. Usually we adopt a commonsense view, a view that considers our knowledge of an object to be valid when *all* of the properties we have "*at*tributed" to the object have been "*con*tributed" by the object. We are right about the redness or roundness of the object when our thoughts about its color or shape have gotten themselves into line with the actual color and shape which the object has somehow contributed to our experience of it. Similarly, we should be right about its *being* an object and the temporality and spatiality of the object when our thoughts about its object-ness, temporality, and spatiality have similarly been contributed to our experience of it by the object—not by us! But if Kant is right that we apply the concept of object (a priori) in the course of constructing any experience we can ever have of it (or any other object), how can we ever apportion a conceptual quality like "object" not to us but to the object itself, the side of the equation that makes the most (common) sense?

The two possible answers to this question are easy to derive, coming as they do from nature and from nurture. Let us start with the answer from the nature side. In the biological language of our day, we might say that the nature argument would look to evolution to answer the question. Here we would say that the concepts our faculty of understanding always brings to the interaction of ourselves with the objects we experience are *genetically determined concepts.* Since this is the current way we express the idea that they are "simply part of our nature," we might reframe Kant's view by postulating that the conceptual form we apply to shape our experience is determined by our genes and the way those genes shape our brains—and hence the way our brains shape our experience.

On this nature-side view, we would appeal to evolution to answer the commonsense question about how objects in the world have contributed to the object concept we apply in constructing every experience we have of that world. Presumably, so the argument goes, as thinking evolved, our ancestors who were right about certain important features of the world (causality, spatiality, temporality, object permanence, and so on) had a huge advantage over their fellow prehumans who were wrong about them. Since the "external" world manifests such pervasive features as causation, time, space, permanent objects, and so on, there was selection pressure toward brains that used these concepts in constructing our ancestors' "internal" experience. On this telling of the story, those prehumans whose genes coded for brains clever enough to get these crucial things right must have had a survival and reproductive advantage over less clever comrades, so these genes became concentrated in the population (and will soon be sequenced by the Human Genome Project).

This evolutionary story seems to solve our problem. True, in each case I do indeed "contribute" the property of object as Kant described, while the object itself contributes only crude sensory information. But the world of things (where we would ultimately like *all* the properties to come from) has "contributed" to that concept I bring to the interaction because my faculty of understanding has itself been shaped by the "real" shape of the world through the forces of evolution and natural selection. I can be sure that I am right about even these conceptual qualities of things because the genes responsible for the neural structures that give rise to them have been sculpted for millions of years by the realities of that very world.

There is, however, another way to say that even our basic concepts have been shaped by the objects of our experience, although we apply them a priori in each case as we interact with those objects. This is the nurture argument, which may also be told as a biological story. In this version, each individual does not have genes preprogrammed for brain structures that apply concepts like "object" to the data of experience. Instead, each individual is born with a brain that is not yet mature, a brain that is shaped by the environment throughout life, or at least throughout early life. Each baby growing up in a spatial, temporal, object-containing world has a brain which is shaped by that world to apply these concepts throughout the rest of his or her life.

Here evolution has once again solved our problem, but in a more subtle way. Evolution has not, according to the nurture view, given rise to a brain that manifests the conceptual apparatus most adaptive to one specific environment. Instead, evolution has developed a more powerful and flexible strategy: the conceptual apparatus that won the natural selection sweepstakes was the one that exhibited maximal plasticity to *any given* environment, not maximal accuracy for a specific environment now long-since past. Brains less good at shaping themselves in early life to whatever features the world had

to offer were brains that would not compete well. Genes programming for this capacity of the brain to mold itself to these pervasive
features of the world are genes that confer tremendous survival advantage. Longer survival would usually lead to greater reproductive
success in the form of more surviving offspring, who themselves
would compete successfully because they carry the adaptive genes
(and *these* will be the ones sequenced by genome researchers).

Before we even begin to consider which of these two stories might
reveal the evolutionary truth of the matter, it is striking to note how
this entire argument has introduced a historical element into what
has thus far been philosophical theorizing. Which way did evolution
in fact unfold? This "historicism" is inherent in any biological story
that hopes to solve philosophical problems, since evolution unfolded, to put it simply, the way it unfolded. Indeed, we must not
forget that evolution is *still* unfolding, and this has implications for
any appeals to it for solutions to basic philosophical questions. In
our nature-side version of the story here, for example, the continuing progress of evolution introduces the possibly unsettling implication that all our presumed knowledge is not valid—or at least *not
yet* valid—either because there has not yet been enough evolutionary time for the world to shape our faculty of understanding to the
precise details of the shape of the world, or else because even these
pervasive features of the world (spatiality, temporality, and so on)
are themselves slowly changing, and so the best we can hope for in
our knowledge is an approximation to a moving target. This "possibly unsettling" consideration would obviously make our nurture-
side version of the story more appealing to anyone who prefers to
feel certain that we are actually right about all our presumed
knowledge.

Appealing as either argument might be, however, we have changed
our standards of evidence by shifting into a biological idiom to try
to solve a philosophical problem. If we are trying to evaluate which
of two biological theories of brain evolution is correct, it is not for
us to apply arguments about which theory is more appealing, more
rational, or more elegant. The standard of evidence in the biological
sciences is the standard of the *facts*, as established through

observation and comparison of historical data or data obtained through controlled scientific experiments.

How would we devise an experiment to settle the question of whether our object concept is inherited through genes that have shaped themselves over the millennia to a world of objects, or whether children, in interacting with a world filled with objects, are nurtured by this environment to take up the object concept and apply it throughout the rest of their lives in constructing their experience? It is unlikely that the Human Genome Project will find a solution to that question, since even the most extreme biological reductionists do not fantasize that there will actually be an "object-concept gene" to be discriminated among those 3 billion A's, C's, G's, and T's. (Our discussion of biology below will show why this would not even make a good fantasy.) It also seems unlikely that fossil remains of our prehuman ancestors will give us much to go on, since it will always remain a matter of some speculation as to whether it was their emerging object concept that conferred their evolutionary advantage, or some other capacity, like tool building or language. No: only one experiment could really do the trick.

All we have to do to settle this whole question is start with two identical twins, born today, whose identical genetic makeup has been shaped by our world, with its permanent objects. We then simply have to keep one twin here and send the other one from the moment of birth off to some other galaxy that does not manifest permanent objects. In this other galaxy, such objects as there are might come and go capriciously. Perhaps in this other galaxy you sometimes reach for a chair and the chair breaks up into a thousand little chairs, and then later forms back into the big chair, like a mercury droplet dispersing and then recoalescing. Since the twin in that galaxy inherited genes that were shaped by our world, we can finally answer our question by waiting some years and seeing what her experience is like. Would her plastic brain shape itself to the real conditions of that world, so that she ultimately constructs her experience using some concept other than object—one more appropriate to her environment—or would she go through life constantly surprised to find that objects are not behaving as they "should"? If genes selected in our

galaxy are indeed directing the shape of those brain structures responsible for such a concept, then we would expect this profound
mismatch to impair permanently her capacity to know her world.

While this may sound like a fantasy, it is exciting indeed to report
here that this experiment has now been successfully carried out. The
results are in and will be discussed in some detail over the next two
chapters.

Before moving on to those results, however, it is important to note
that either way the experiment could have turned out, the very possibility of framing the question as we did already sheds some light on
the subject. Once we accept the Kantian notion that our mind actively applies concepts in the course of constructing our experience
of the world, it is easy to become preoccupied (as both Kant and
Freud did) with the way the mind's own structure dictates all that
we come to know of the world. With its appealing advance over Descartes's view of the mind as merely a thought-containing object, it is
natural to look instead to this more active view of the mind's participation in the world it knows. But whether the nature side or the
nurture side answers the question of *how* the world of objects can
contribute to the basic object concept we (actively) mix with our
world, in asking this very question we have begun to consider the
world's participation in the mind (Hegel) and not just the mind's participation in the world (Kant).

In the end it is only this mutual participation that can bring
thoughts and things into a relationship of sufficient intimacy that
philosophy can be saved from the independence of the two that
caused Descartes to turn to God for guarantees, and caused Kant to
accept the existence of an unbridgeable schism between the world-
as-we-know-it and the world-as-it-really-is. As we examine all the
evidence that results from our experiment, we are thus searching
even a basic, a priori concept like *object* for its hidden signature of
Will's "helping hand of things" and learning, as he put it, "how we
think with the help of things."

Put another way, our experiment has been designed to answer an
instructive (if counterintuitive) question: How does the a posteriori
world of things contribute to the a priori concepts we use in

constructing our experience of that world? By our original definitions, the "a priori" should be free from such empirical contingencies; the situation has already grown more complicated, however. In
our experiment, if the world as biologists conceive of it did not manifest permanent objects, then our assumption is that we would not
apply that a priori concept—either because some other more adaptive concept would have evolved in that world as a permanent feature
of our mental apparatus (the nature result) or because we each
would be growing up with our concepts shaped by that other world
(the nurture result). In either case, we have complicated our former
Kantian distinction where the reality of the world itself contributed
only sensory content while we contribute the form of any world we
could know. Whether through brain evolution or early postnatal
brain maturation (or both), the form as well as the content of our
experience of the world may well be contributed by that world, as
our commonsense view would demand—even if it might have
wished for some more straightforward mechanism for contributing
on the conceptual side.

Something like this line of thinking was precisely what led Hegel
to suspect the "purity" of Kant's faculty of reason. Kant knew that
understanding's concepts had to be contaminated by the a posteriori
in always being mixed with the (contingent) sensory contents of sensibility when they do their job. He was able to see reason as "pure"
because it involved the manipulation of these a priori concepts in
isolation from the a posteriori world; we might say that reason
pulled the strings, but it never touched the puppets. Our more complicated view now undermines the "purity" of reason by revealing
traces of that world hidden within the concepts themselves, and so
leaves none of our mental faculties free from the influence of reality.
When Hegel argued that reason is not relative only to our consciousness but to reality itself, this was the point he was making. (Think
for a moment about hand puppets rather than marionettes.)

Before we continue our exploration of the relative contributions
of nature and nurture to our knowledge of the world, it is worth
noting that most of what has been said thus far about knowledge
could equally well be said about ethics, another field that we will

look at closely later. We have identified a dialectic by which *concepts* in our minds shape our knowledge of the world just as the world itself contributes to the shape of those very concepts. We could similarly talk about the way *values* in our minds shape certain social practices in the world just as those practices contribute to the shape of our values. Once again, questions of nature and nurture arise. We might wonder whether the social practices (including customs of childrearing, sexual mores, religious institutions, and the like) "nurture" young children to take up the values they will hold as adults, even as the shape of some possibly "natural" inborn values might have given shape to those practices in the first place (a possibility more appealing as the practices in question become more universal).

If we had started with the field of ethics, we would have needed a very different twin experiment to investigate the contributions of nature and nurture to our values and social practices. These experiments would have been trickier than the one we did. We might have transported one twin to a galaxy where people never keep their promises to see if she would continually be disappointed by these broken promises or whether the value she would attach to promise-keeping would be shaped by that "ethic" rather than by our planet's long history of promise-keeping as adaptive behavior. We could similarly try sending one twin to a galaxy where no value is attached to truth-telling.

These experiments would be even trickier than the one we did because it is hard to imagine that the practice of making promises would continue in a galaxy where none are kept, and so there would likely be no promises, broken or otherwise, with which to test our subject. Similarly, it is hard to see how useful communication could possibly take place without at least some shared assumption that people generally mean what they say, so it is not clear how constant lying could be maintained.

But such details are of no concern at present. The only point is that, as we now turn back to our philosophical explorations of knowledge, we should keep in mind the possible relevance of whatever we discover to other, very different areas of human experience, including psychology and ethics.

6

How Quickly They Grow Up

Although Kant first realized that we apply concepts like "object" a priori to our experience as we construct that experience from our sensory impressions of things in the world, our twin experiment has been designed along more Hegelian lines. We seek to uncover in this a priori object concept the hidden signature of the helping hand of those (a posteriori) objects in the world. In the biological idiom of our day, a real world of objects could write this feature into our conceptual grid either through some genetically inherited endowment or through some shaping of our brains by the environment early in life (or some combination of the two). In our definitive experiment, we took one identical twin and sent her to a galaxy that does not manifest permanent objects and kept the other twin here on earth as a control.

Any good experiment needs a control. The results here on earth turn out to be almost as interesting and complicated as the results in the other galaxy, so let us begin here on earth. One reason for the complexity is that our control twin is unable to talk to us for a year or two about what she is experiencing. Through careful observation, we can feel relatively certain that by the age of two years she is attaching names to objects like "ball," "dog," "mommy," and "daddy" as if she appreciates that they are relatively permanently existing objects. She applies these words correctly even when the object in question is temporarily out of sight, running in the yard, wearing a new hair style, or returning from a day at work.

Object is indeed only one of a number of basic concepts already part of our control twin's repertoire by two years of age. When the first pioneer of this sort of experiment began to study "normal" children like our control twin, Jean Piaget found that two-year-olds apply such concepts as object, space, time, and causation as they organize their toddler's view of things—the very concepts Kant considered to be "necessary conditions of any possible experience."

But what about experience in the first year or eighteen months of life? Since newborn babies cannot use words to tell us about their experience of the world, we have to use a certain amount of inference to collect the relevant data even from our control twin. Furthermore, unfortunately for our experiment, it is the experiences of those first months and year on which our entire result depends, since any "nurturing" the world of objects might offer our child's object concept might already have been completed before she begins to talk. In order to discover whether some genetic endowment provided her with an object concept from birth, we need to frame hypotheses about her view of the world and then test these hypotheses by observing her behavior and reactions under various conditions.

When Piaget did just that, he discovered that in all likelihood babies come into the world with none of these concepts that they possess at age two. This left him with the monumental task of discovering how such concepts come into being, and this gave rise to his familiar "stage theory" of cognitive development.

Jean Piaget, one of the most prolific writers of this century, was born in Neuchâtel, Switzerland, in 1896 and died in 1980. He is often described as a "psychologist," but that is only because this term comes closest to any profession that existed when he began his work. Since the philosophical study of knowledge is called "epistemology," Piaget dubbed the new discipline he founded *genetic epistemology*, which was concisely described by one of his students as "an experimental philosophy which seeks to answer epistemological questions [questions about how we 'know'] through the developmental study of the child." Piaget's genetic epistemology is grounded firmly in biology, as we might expect from someone who started his career as a zoologist and earned his doctorate at age 22 for a thesis on the classification of molluscs! Just one decade later, addressing himself to "the traditional problems of the theory of knowledge," Piaget wrote:

> The problems we are about to study are *biological problems*. Reality, such as our science imagines and postulates, is what the biologists call environment. The child's intelligence and activity, on the other hand, are the fruit of organic life (interest, movement, imitation . . .). *The problem of the relation between thought and things, once it has been narrowed down in this way, becomes the problem of the relation of an organism to its environment.* [emphasis added]

In his premier study of infancy, *The Origins of Intelligence in Children* (1936), Piaget attempted to describe how the concepts of object, space, time, and causation develop in young children over a series of stages, each of which builds upon the preceding stage. Starting with only primitive reflexes such as sucking and grasping, the infant applies these reflexes repeatedly—soon refining them and using them to differentiate various features of the environment. Although the newborn sucks at anything that comes near her face at birth, by the second week of life she has already become expert at finding the nipple and differentiating it from the surrounding tissues, and some weeks later she coordinates the movement of her arms to suck her thumb systematically for soothing. Later still, she can grasp objects to bring them to her mouth and explore them through this same sucking mechanism.

To say that two-week-old infants become "expert" (at finding the nipple) is to view even their earliest sensory and motor actions as a nascent form of intelligence. One of Piaget's biggest breakthroughs in his study of cognitive development was his recognition that in the earliest "sensorimotor stage" from birth to 18–24 months, the same process of cognitive development has already begun that will continue throughout the rest of childhood and even into adult life. Piaget used the term *schema* for any broadly defined cognitive structure that we use to organize our world, and so he spoke of the schema of sucking, the schema of sight, the schema of prehension (grasping), and so on. Piaget recognized that even these most primitive schemas are used by the infant to assimilate the environment—to "take in" features of the environment in the same way our digestive system takes in nutrients.

Piaget's idea was that in repeatedly applying a schema (such as sucking or grasping) to the world, the infant both generalizes that schema and differentiates it. Thus, sucking becomes generalized as at first only fortuitously placed objects are suckable, then fingers, and soon everything in the child's world. But with repeated use, the schema of sucking also undergoes differentiation as the infant soon discriminates objects to be sucked when one is hungry from those to be sucked when one merely wants comforting and is not hungry, from those to be sucked out of curiosity and exploration.

By using the word "assimilation" to describe the process by which even primitive schemas are used by the infant to organize her world, Piaget is already painting a Kantian picture of the mind as a grid of schemas—now sensorimotor schemas, such as sucking and grasping, but soon to evolve into conceptual schemas, such as object, space, and time. Indeed, Piaget believed that conceptual schemas such as object, space, and time *develop from* more primitive sensorimotor schemas such as sucking, grasping, and seeing. With further generalization and differentiation of each of these more primitive schemas, the infant begins to appreciate that the rattle she sucks is the rattle she sees and the rattle she grasps. When *one and the same* rattle emerges as a thing capable of being assimilated by looking, grasping, sucking, listening, and so forth, the child's network of intercoordinated sensorimotor schemas begins to give rise to a primitive "object concept" that itself continues to mature through a series of substages until the time that language appears and names get attached to these objects.

Although the two-year-old emerges from Piaget's sensorimotor stage of development with concepts such as object, such "sensorimotor intelligence" as she has is still quite primitive. During Piaget's "pre-operational stage" (which lasts from 18–24 months to 7–8 years), the child again goes through a process of repeatedly applying her new conceptual schemas to the world and again both generalizes and differentiates these through repeated interaction with that world. Concepts like object thereby mature to include all those features we adults seem to take for granted.

Perhaps the most obvious of these features is the conservation of the mass in those objects, as demonstrated in Piaget's famous experiments with clay. When children four or five years old are shown a ball of soft clay which is repeatedly rolled into a long thin snake and then back into a ball, they will, upon questioning, either say that the ball has less clay (if they focus on the snake's length) or more clay (if they focus on the snake's thickness), but they will almost always choose one or the other, exclusively. By about the age of five or six, most children can pay attention to both length and thickness, but not simultaneously, and will alternate between saying of the snake "there is more" (since it is long) and "there is less" (since it is thin). Only by about the age of seven, when the preoperational stage gives rise to "concrete operations," will children coordinate both of these perceptually based judgments and realize they exactly compensate for each other at all times, and that mass is conserved.

The same can be seen when presenting a four-year-old with two rows of small stones. Although both rows have the same number of stones, the child will tend to say there are more stones in one row if it is stretched to be longer than the other, and will often persist in saying this even as she watches more and more stones being squeezed into the shorter, denser row. At about the age of seven, however, the child rather suddenly thinks that your questions are quite stupid and says, "Of course there is the same number, nothing was added or taken away." Just as children playing the first game suddenly begin to operate as if the conservation of mass is obvious to them, children playing the second game now identify numbers of things as a feature that the world manifests. Piaget would say that the child is now applying the schema of mass conservation or the schema of number to her experience of the world—that these concepts are part of the grid she uses to assimilate the world in a way quite consistent with our original Kantian model.

But we can see that there is something very different here from

Kant's original notion of these concepts as a priori. True, the seven-year-old seems to be applying the concept of number *to* her experience: she thinks the question is stupid because she finds that "numbers of things" are there to be found *in the world* (having been applied by her mind in constructing any experiences she has of that world). Yet, Piaget's story could only be possible in a world that does in fact manifest conservation of mass and relatively permanent numbers of things. Whether the child's brain is simply maturing over these months and years along the genetically preprogrammed path that emerged from evolution in a world manifesting such constancies (the nature argument) or whether the child's plastic brain is shaping itself to these constancies of the environment (the nurture argument), Piaget's discovery of the development of our a priori concepts leaves room for the "contributions of the a posteriori world of things" to these a priori concepts in just the muddied way Hegel suggested.

In studying the process by which our evolving assimilatory schemas form the adult Kantian grid that we throw over the world as we construct our experience of it, Piaget thus discovered that, in a series of fairly predictable stages, these grids themselves develop with our continued experience of the world. He used the term *accommodation* to describe this mechanism by which our conceptual grids change at each of these stages of development to "accommodate themselves" to the realities of the world. Once these cognitive structures have accommodated themselves to such newly recognized features of the world as object, space, mass conservation, or number, they then become the new assimilatory schemas that become applied "a priori" and are once again, through repetition, generalized and differentiated as thinking becomes more sophisticated and abstract. (The concrete operational stage of "numbers of things" gives rise for many people during adolescence to a stage of formal operations, where algebraic symbols can represent quantitative notions without the need for numbers of stones or even counting on fingers!)

Piaget's observations throw an interesting light on one of the fundamental nature-nurture debates in philosophy: whether it is really possible to "learn" basic concepts, since learning by its very nature presupposes the application of some concepts. This debate over whether basic concepts such as number or object can be learned has—before Piaget's empirical studies—left philosophers in some very awkward positions. On the one side, philosophers like Isaac Newton's German rival, G. W. Leibniz, were forced to posit dormant congenital concepts (we might say genetically programmed, but in need of an environmental "releaser" or "inducer"). On the other side, philosophers like our British empiricist John Locke were forced to posit creative ways the mind might "learn" such basic concepts.

We can now instead look to the emotional experience of our seven-year-old for direction. The seven-year-old did not seem to be proud of her apparent "discovery" of conservation of mass or numbers of things as she might have been for other things she learned: she suddenly thought the whole question was *stupid*. If philosophers have obsessed over whether it is possible to learn concepts such as number, we can now see how such concepts are acquired through a process we might want to distinguish from "learning" in the traditional sense—the process Piaget called the accommodation of our conceptual schemas to the realities of the world. Whether we want to call this process of accommodation a "special kind of learning" or "a way of acquiring concepts other than by learning," it does provide a way to understand how, through stages of cognitive development, *our experience with objects enables us to "objectify them."* Although most of his extensive research concerned the mechanisms of what he called assimilation, Piaget's articulation of this reciprocal process of accommodation was, in a sense, simply a continuation of the Hegelian philosophical program to search for ways we "think with the help of things."

Although we have thus far been talking only about our control twin here on earth, it is worth remembering that our experiment is designed to sort out the contributions to our developing basic concepts between nature and nurture. By reframing the problem of the relationship between thoughts and things in terms of the relationship of an organism to its environment, Piaget the biologist has reminded us how complicated this relationship really is—because of how complicated human development really is. From Piaget's perspective, the process which leads to the development of our object concept may well be controlled by genes—but by genes that give rise to instinctual *behaviors* like sucking, grasping, and looking! On this version of the story, our genome preprograms structures in our brains, mouths, hands, and eyes that ensure that any objects a particular environment provides will be experienced through enough different sensorimotor modalities that these can be coordinated by some less preprogrammed (more "plastic") part of the brain into a preoperational assimilatory grid that will include the object concept. Examples of genetically controlled behaviors are well documented in the animal world, and it is most definitely not a fantasy to think that some of the genes sequenced in the Human Genome Project may some day be recognized as the genes responsible for instincts like sucking, grasping, and looking.

Piaget was not alone in his observation that cognitive and other human capacities evolve through a series of stages of development, and there is much evidence to support this idea both from nonhuman primate studies and from human observations (or just ask any parent or pediatrician). Freud's model of the mind that we saw earlier also comes with a series of stages that ultimately give rise to the final shape of those mental structures, and Freud's oral, anal, phallic, oedipal, latency, and other stages come with parallel psychological stories to Piaget's notion of "coordinating the schemas of sucking, grasping, and seeing." As we have seen repeatedly, each thinker's version of the story reveals his other agendas and ulterior motives in developing such a model of the mind in the first place, Piaget's being the evolution of Kantian concepts and Freud's being the development of psychosexual capacities and symptoms.

These other agendas and ulterior motives led these thinkers to focus on infancy, childhood, and adolescence, with little attention to possible stages of adult development. The psychologist Eric Erikson more recently described yet another series of stages of development which continue throughout adult life in part because his agenda focused on the relationship between the individual and society—a point to which we shall return in the second half of this book. Indeed, when we remember from the end of the previous chapter

Stages of Development: Four Models

Age	Piaget's model of cognitive development	Freud's model of psychosexual development	Erikson's model of identity development	Kohlberg's model of moral development
Infant	Sensorimotor intelligence	Oral stage	Trust vs. mistrust	Heteronomous morality
Toddler	Preoperational intelligence	Anal stage	Autonomy vs. shame, doubt	Individualism, instrumental purpose, and exchange
Preschool age	—	Phallic stage (start of Oedipal complex)	Initiative vs. guilt	Mutual interpersonal expectations and conformity
Primary school age	Concrete operational intelligence	Latency	Industry vs. inferiority	Social system and conscience
Adolescent	Formal operational intelligence	Adolescent sexual development	Identity vs. identity confusion	Social contract or utility and individual rights
Young adult	—	—	Intimacy vs. isolation	Universal ethical principles
Middle age	—	—	Generativity vs. stagnation	—
Old age	—	—	Integrity vs. despair	—

Although each of these models has been criticized for one or another specific feature, both these and the models used to criticize them all recognize the presence of some identifiable stages of transition and growth as a conspicuous feature of human development. There is a wonderful irony that the stages proposed by Freud, who treated adults, are completed at adolescence, while those of Erikson, a child psychologist, continue through adulthood to old age.

that all our considerations thus far might apply equally to other areas such as ethics, it becomes very important to include the whole life cycle and not to over-focus on the earliest stages of development. In writing about stages of moral development, thinkers such as Lawrence Kohlberg and Carol Gilligan—while claiming to present contrasting views of how moral development proceeds—each elaborate models that are consistent with the possibility of continued moral development into adult life. As we shall see in Chapter 11, our current understanding of the human brain is very consistent with this notion that continued cognitive and moral development is possible into adult life.

When it comes to basic cognitive and conceptual concepts such as object, however, it is natural to focus on infancy and childhood. The two-year-old, after all, already manifests primitive concepts of object, space, time, and so on, and so the development of such concepts must begin from a very early age. Having used our control twin here on earth to introduce the notions of assimilation and accommodation of cognitive schemas through a series of stages, we are now in a position to look at what has been happening to the experimental twin we shipped off at birth to a totally different kind of environment.

7

Feelings and Things

In founding the field of genetic epistemology, Jean Piaget reminded us that when we talk about the relationship between thoughts and things, we are merely talking about one aspect of the relationship of the human organism to the environment. As a biologist, Piaget recognized that thinking is simply one among many ways that we humans adapt to our environment. He therefore reasoned that there ought to be some similarity between how thinking works and how all of our other adaptive mechanisms work. In looking at such a basic adaptive mechanism as eating, Piaget distinguished the way in which we change the shape of food as we *assimilate* the food to ourselves (by cutting, chewing, and digesting it) from the mechanism by which we *accommodate* the shape of ourselves to the food in the process of that same adaptation (by opening and closing our mouths, swallowing, and even adjusting the digestive juices we use to the specific physical and chemical properties of the food in question). In searching for the application of these same mechanisms of assimilation and accommodation in the mode of adaptation to the environment we call thinking, Piaget was similarly able to find the ways in which our schemas assimilate reality to whatever conceptual grid we apply even as the shape of that grid accommodates itself to reality through a sequence of stages. By taking questions about the relationship between thoughts and things and placing them into the biological context of the relationship between organism and environment, Piaget provided for the possibility of empirical research on basic questions in the philosophy of knowledge. He brought cold, abstract concepts to the warm-blooded realities of human existence.

But Piaget always set the conditions of his experiments so that the children were under the minimum possible stress—they were always well fed, napped, and so on—and in fact he has mostly been criticized for not being warm-blooded *enough*. After all, the relationship in question is not that of a human infant interacting ..... pieces of clay and stones in lines, but of a baby interacting with its mother and father. The entire discussion of "objects" takes on a new and very different light when we remember that the objects of greatest interest to these little human organisms are other human organisms! If Piaget reminded us that the relationship between thoughts and things is just a part of the relationship between organism and environment, we might now remind ourselves that the organism in question is a young child, and, in the words of the psychoanalyst-philosopher Arnold Modell, "For the young child, the mother [in its generic sense of 'caregiver'] and the environment are indeed synonymous"—either literally when being held or fed, or else more indirectly through the environment provided and maintained by her even when she is temporarily elsewhere.

Since the use of words carries with it some assumption of the existence of the objects to which the words refer, it is perhaps impossible to describe in words what the experience of an infant might be like before the development of the object concept. But if we cannot find adequate words to describe the "thoughts" of a newborn baby, we can perhaps at least identify with the *feelings* of a baby whose bodily needs were all well attended to in the womb and who now, after birth, suddenly finds herself with repeated experiences of thirst, hunger, and discomfort. Rather than receiving her food passively through the umbilical cord in a temperature and pressure-controlled cabin, the infant is suddenly forced into a more active role. Her hunger makes her cry, and the crying brings relief in the form of milk. Although it is questionable whether the newborn appreciates that this milk is an "external object" that exists outside herself, we can at least appreciate the feelings in the nursery when the uncomfortable infant is hungry, wet, and cold and her crying leads to relief in the form of warm milk and a warm, dry diaper.

This view of the relationship between organism and environment helps us understand why the insightful British pediatrician and child psychoanalyst Donald Winnicott insisted that there is "never just an infant." He did not simply mean that someone has to take care of this small helpless creature. He meant that any baby who survives comes "attached to" at least one other person, and without such attachment (like the earlier attachment to the uterine wall), life itself is impossible for human beings. With this perspective, the developmental psychologist Robert Kegan has reminded us that the reflexes which Piaget called to our attention have an unmistakable evolutionary message in helping the infant secure this life-supporting attachment. The new baby's grasp of the mother's garment, her orientation to the mother's eyes, and even her sucking on the mother's nipple and finger not only assimilate and differentiate the environment but seduce and secure the mother's attachment through this natural beguiling.

Psychiatric explorers such as Modell, Winnicott, and Kegan focus our attention on this drama of the nursery to help us understand the answer to a subtle question in the philosophy of knowledge: here is a baby who will begin to appreciate reality, but *as distinct from what?* From out of what subjective experience does the objectivity of the world emerge?

Remember the situation. The infant is hungry, wet, and cold. She cries. Relief comes in the form of warm milk and a dry diaper. The infant is comfortable again and sleeps. If her need for the milk in fact produces the milk she needs (as it does—if indirectly), why should such an infant come to believe that the milk is part of a separate objective reality and not simply part of a magical world shaped by the infant's own wishes? Perhaps if every single wish were instantly fulfilled, the child would *not* have any reason ever to believe in a separate external objective reality. Perhaps if the "perfect mother" could always instantly satisfy the infant's needs and desires, the child would come to believe in her own omnipotence and ability magically to make appear any object that is wished for—a variation of the galaxy to which we have sent our experimental twin. (If such a "perfect mother" could actually make this happen, the grown child would presumably have one word for both an object and her need for it, and the linguistic researcher who discovered such language really would make a contribution to philosophy!)

Of course in the real world, no mother can be perfect in that sense. Inevitably, she arrives five minutes later than wished for; and in those five cold, wet, hungry minutes the infant begins to discover that there is indeed a difference between the subjective mother of her fantasies and the objective mother in the external world. Piaget's sterile picture of objects such as toys and clay disappearing and reappearing thus gives way to an experience of sequences of frustration and gratification that are inseparable from the appearance and disappearance of the most important object of all: the "mother." While no mother can possibly gratify instantly every infantile need and wish, what Winnicott called the "good enough mother" can provide enough satisfaction along with the inevitable frustration to provide both adequate nurturing and the appreciation of a reality that is distinct from the child's own fantasy world. It is, in other words, the magical mother of the infant's fantasies who anticipates every need and wish who must eventually be distinguished from the real (and with luck, good enough) mother, as the child begins to appreciate reality.

In that frustrating period of *waiting* for the *mother* to *reappear*, we find the infant confronted with a temporal, spatial reality in which identity-preserving objects prove themselves permanent through time. This is why Modell concluded in his brilliant *Object Love and Reality* (1968) that "the acceptance of painful reality rests upon the same ego structures that permit the acceptance of the separateness of objects." Researchers extending Piaget's methodology to study the development of basic concepts in the mentally ill have similarly concluded that "sequences of frustration and gratification are primary experiences considered to facilitate the differentiation between self and nonself and a sense of continuity and permanence of objects."

Piaget recognized the importance of this view when he noted that the development of our cognitive schemas depends upon the social environment, but we now have a much richer picture of our mental life when we consider that the construction of *external* reality is tied to the acceptance of *painful* reality. We also begin to question whether it is possible to abstract, as Piaget did, our cognitive development from our emotional development. Indeed, the psychoanalyst and researcher S. M. Bell has shown that the quality of infant-mother attachment affects the rate of development of the object concept! Such findings take us a long way from our more sterile Kantian view of fixed concepts, even if Piaget has shown that Kant's fixed concepts develop in us over time. As we begin to see how our love attachments are intimately connected to the acquisition of even our most basic concepts, we start to get a clearer understanding of what was meant above by the "participation of the world in the mind." Here, we also anticipate our later neurobiological understanding that the same structures of the brain that enable us to think also enable us to feel, as Modell concludes in the following passage from his psychological and conceptual study:

> The structures of the mind that determine the individual's relation to reality and his capacity to test reality—his capacity to distinguish between wish and perception—are inextricably bound to both the quality of his earliest human love relationships and to the gratification that is afforded him by human beings in the present. *The capacity to know and the capacity to love are not . . . entirely separate functions.* [emphasis in original]

With this background, we can now reveal the results of our twin experiment. Since the objects of greatest interest are not rattles and pieces of clay but mothers and fathers, we actually have infants who grow up in that other galaxy—right here on earth. When the mother is alcoholic or the father is manic-depressive and is repeatedly hospitalized, infants do indeed grow up in a world where "objects come and go capriciously." By making empathic efforts to appreciate the conceptual grid applied by adults who had such unfortunate upbringings, psychiatrists find that these patients seem to be applying a concept of "object" that is very different from the one usually employed, especially with respect to the permanence and constancy of those objects. Although these adults act in some ways as if they know that most inanimate objects are permanent (they do not bump into walls), they do not appear to possess the concept that objects continue to exist once out of sight when these are objects they care most about (some "precious" inanimate objects, but more often the people in their lives).

These observations are also supported by psychoanalytic researchers who have studied the impact on child cognitive development of harsh institutionalized environments, such as the crowded orphanages of London created by World War II. Infants in these environments typically get minimally adequate nourishment, but they often get something far short of the interpersonal "good enough mothering" that appears to nurture the development of Kant's object concept through a balanced mix of sequences of frustration and gratification (the latter being absent). These dedicated researchers—including Freud's daughter Anna, along with Rene Spitz, John Bowlby, and others—have left unequivocal the impact of the environment on even our most basic concepts about ourselves and the world. At the extreme, we also have the thirteenth-century "experiment" of King Frederick II—described in the opening pages of this book—which concluded that a mother's loving words are as necessary to survival itself as food, shelter, and clothing.

The results of our experiment obviously lend support for the nurture argument, but with two added twists we had not built into the experimental design. For one, we have been reminded that genes for instincts like sucking and grasping may have almost as much to do with the development of our object concept as the hypothesized "object concept gene," which has to be more or less ruled out by the experiment. For the other, we have not only come a long way from Kant's fixed, rigid view of concepts when we see how they can develop through our interactions with the world and even become impaired in certain circumstances; we have also discovered that the cuddling of infants and children is important not only to their emotional development (as one would assume) but to the development of the faculty of understanding first described in such bloodless terms by Kant. We must not forget that knowing and loving do indeed rest on the same ego structures (or brain structures) as we continue to broaden our perspective on human experience.

What we now appreciate is that the actual environment in which the developing infant first approaches the "external world of objects" is an environment that not only contains "things" but also contains *people*. In other words, the environment in which subjectivity (the infant) first approaches objectivity (the world) is in fact another subjectivity (the mother)! This is perhaps just a more complicated version of Piaget's own claim that cognitive development depends on the social environment, but this very warm-blooded view does raise a fundamental question about the objective status of Kant's basic concepts—his categories of the faculty of understanding. We have now replaced Piaget's (and Kant's) picture of the *individual's* subjective grid organizing our understanding of the "world-out-there" with Winnicott's picture of an infant-mother dyad. From this view that there is "never just an infant," neither is there such a thing as "just a subjectivity," but only what might be called the *intersubjectivity* of the collective experience in the nursery. The term "intersubjectivity" tries to capture the social aspect of even individual subjective experience; we may need to broaden Fred Will's challenge to understand how "we think with the help of things" to explore how we "think with the help of things *and other thinkers*." This will be especially true when we turn from the philosophy of knowledge to related areas such as ethics and the relationship between moral values and social practices. Indeed, there one might question whether it makes sense at all to talk about a "moral value" or a "social practice" when considering only one individual's experience, in the way that philosophers and psychologists have traditionally done.

But before temporarily leaving philosophy to explore some of these other areas, there is one more view of the relationship between thoughts and things (be it conceptualized as organism and environment, or infant and mother) that we have yet to consider. This is the not-so-mysterious fact that sometimes our concepts (ideas) about not only how the world is but how we think it *should be* leads to changes in how the world "really" is.

In the tradition of many a philosopher, I can ponder for hours the relationships between my thoughts about the trees I can see through the window and those trees "in themselves" out there in my yard. But we should not get too far in the pondering without taking note that *I planted those trees.* All right, to be technical, I hired a landscaper to plant them. But the fact remains that the relationship of greatest consequence between my thoughts and those things is how my thoughts led to their being there. Since much of what is real was made by humans, there is clearly a deeper sense in which we construct reality, both in terms of the toys and pieces of clay we come to know and the social practices and institutions that interact with our value systems.

Part Two

There must be ways to put the mind back into nature that are concordant with how it got there in the first place.

Gerald M. Edelman

8

The Anthropic Principle

Having begun our nature-nurture odyssey with basic questions about knowledge, we have been led by Western philosophy to focus our attention on Descartes's original distinction between thoughts and things. Since our thoughts about things arise from an interaction between them and ourselves, all of our considerations thus far have centered on the dialectical relationship between those things "out there in the world" and our thoughts about them.

At first we focused on how the structure of thought itself shapes our experience of those things. In the spirit of Kant and Freud we reminded ourselves that the structure of the mind (or brain) conditions much of our experience of the world. If we speak of the "contributions" that thoughts and things make to one another, we might say that Kant recognized that things contribute crude sensory information such as colors, shapes, and motions to our thoughts about them, but he emphasized the opposite direction. That is, Kant emphasized the contributions of the structure of thought itself to our everyday experience of things. In highlighting for us the process through which we actively construct the shape of all of our thoughts, Kant elaborated his famous categories of the faculty of understanding, which set conceptual boundaries on any experience we can ever have of the world.

No one would doubt the importance of the contributions made

by the shape of our mental or neural structures to any experience we have of the world. We can easily imagine, for example, two individuals walking into the same party; their sensory organs receive the same input of shapes, colors, motions, and so forth. One individual feels warm and happy in seeing all of his friends in the room, and gladly joins in the party. The second individual has an agoraphobic panic attack when he sees the crowded room, becomes tremulous and sweaty, and beats a hasty retreat. In making (common) sense of the dramatic difference between these two experiences, we take for granted Kant's constructive view of experience: the two people applied different Kantian grids in constructing their unique experience from the identical sensory data offered by the party (where here we might complement Kant's conceptual grid with additional *affective*—emotional—categories to account better for how our experience is "mooded").

We were reminded in Chapter 4, however, that the relationship between thoughts and things is a two-way street. In criticizing Kant's view of the mind as a steel filing cabinet with fixed conceptual labels on each drawer, Hegel focused our attention on the opposite direction: the contributions of things to our thoughts about them. Here Hegel was not merely talking about the contributions of such crude sensory information as shapes, colors, and motions but the more subtle ways in which things in the world contribute to the shape of the conceptual grid we apply in constructing our experience of them. In dissecting his experience with a more holistic microscope than Kant's, Hegel discovered that our faculty of reason is capable of discovering discrepancies between the conceptual features of the objects in the world and our knowledge of them. In his masterpiece, *The Phenomenology of Spirit*, Hegel insisted—in one of philosophy's boldest moves—that when such a discrepancy is discovered by reason, then "consciousness must alter its knowledge to make it conform to its object." While Kant's view of knowledge as an activity had surpassed Descartes's more passive view of knowledge, this insight enabled Hegel to take yet a further step, making knowledge not merely an activity but an achievement.

Hegel's idea of knowledge as an achievement was actually far more radical than has been presented thus far. In order to understand his position, we have to expand the two-way street to make room for what counterintuitively might be thought of as a *third direction* between thoughts and things. True, the structure of our thoughts contributes to our experience of things. And true also, the nature of things contributes to the structure of thought itself (as both our twins showed us). But, as we noted at the end of the last chapter, thoughts also contribute to many things by actually *creating those things.* No philosophical discussion of the relationships between thoughts and things is therefore complete without paying some practical attention to the ways in which at least some human thoughts literally give rise to the reality of some things in the world.

Let us consider a typical example. We have seen how beginning students of philosophy in the Cartesian tradition have annually been asked to ponder the justification for their beliefs about the tables and chairs in their classrooms and dormitories. They soon recognize that the tables and chairs in question contribute crude sensory data pertaining to certain specific characteristics of shape, color, and so forth. They soon also recognize that their concepts about the way the world is carved into objects contributes to their experience of those tables and chairs. And they may further recognize the likelihood that other tables and chairs with which they interacted from an early age gave rise to the conceptual framework into which those specific sensory features are now being organized. Particularly with help from their friends studying anthropology, they may thus recognize that the concept of "chair" or "table" is contributed by the nurturing of a world manifesting such objects. Indeed, if their studies take them as far as Hegel, they may recognize that even their experience of the tables and chairs as relatively permanent objects existing in space and time is supported through subtle contributions of other objects in the world (including houses, trees, and so on) that have helped shape the basic conceptual grid they are now applying in constructing their experience of the world.

In contemplating the relationship between thoughts and things in this way, it is striking that most students are more willing to introduce the contributions of other tables, chairs, and similar objects (for example, from their childhood) than the contributions of *other thinkers*. But the simple truth is that the most basic relationship between thoughts and those things called tables and chairs in their classrooms and dormitories is that some people had the thought that, if students were going to learn, they would need tables and chairs in their classrooms and dormitories at which they could study. And so these other people—perhaps long since dead—ordered the tables and chairs from some manufacturing firm for the school. If any thoughts are relevant to our consideration of the relationships between thoughts and things, it has to be the thoughts that *actually* gave rise to those things. Similarly, the thoughts of the entrepreneur who decided to start a furniture factory to capitalize on the thoughts of the educators designing the school likewise have their hidden signatures scrawled all over the tables and chairs about which the students are cogitating.

Another take on how we *actually* "construct" reality was captured by the great Oxford philosopher J. L. Austin (1911–1960), who developed the notion of "performative" acts of language. While some things we say describe things or ask about things, other things we say actually *do* things. A possibly old-fashioned but familiar example is the utterance "I pronounce you man and wife," which, when spoken by the proper authority, brings a state of matrimony into existence. Similarly, when my parents said, "We'll call him Edward Mark," this *became* my name. Performative language contributes to our reality every time we sign a legal contract or call the batter "out." Unlike the subjectivist umpire who says, "There are balls and strikes, and I call 'em as I see 'em!" or the objectivist umpire who says, "There are balls and strikes, and I call 'em as they are!" the performative umpire more astutely says, "There *are* no balls or strikes until *I* call 'em!"

We saw that Hegel rejected Kant's idea of a pure reason uncontaminated by the actual physical world. We can now see that Hegel was in fact taking an *extremely* holistic approach to the problem, and by insisting that reason itself is relative to reality, Hegel was acutely aware that reality itself exists in a dialectical relationship with *us!* So long as thoughts sometimes give rise to things in the physical world, we cannot be so self-centered as were both Descartes and Kant in thinking that it is only *my* thoughts that are relevant to this dialectical interaction between thoughts and things. To get a complete understanding of the situation, we must take a much broader perspective and consider the relationship of *thought in general* and the whole *history of thoughts* as they interact with these things in our lives.

But how much of "the world out there" must be considered "relative to" our human reason? True, we city-dwellers spend an inordinate amount of time interacting with obviously man-made things. But as we move from particular tables and chairs to large forests to the great mountain ranges and eventually to the cosmos, there presumably exists some kind of continuum with respect to *how much* human thought is connected to the actual reality of things. Of course, those anthropology colleagues mentioned above will quickly remind us that differing human experiences of the great mountain ranges and of the cosmos are themselves very different: Westerners know Mount Everest as the world's tallest mountain and a great achievement to ascend, while locals know the same mountain as Chumalungma, the Earth-Mother Goddess. Whether a hard-nosed natural scientist could ever coherently and consistently expose such differences as merely romantic and not relevant to philosophical questions about the objectivity of knowledge continues to be an open question (to which we shall return shortly).

But surely the ultimate physical laws and properties of the cosmos are in no way dependent upon how we look at things. When physicists attempt to describe the laws and properties of nature, these scientists simply call 'em as they are: the universe's laws and properties have nothing to do with the physicists looking for them, according to the traditional view. Many a theologian has made this commonsense argument. If anything sets limits on the nature of the universe, it certainly is not us (so it must be God).

This commonsense view that the physical nature of the cosmos is in no way dependent upon our coming to know it has been called seriously into question over the last thirty years or so. In considering the problem of whether anything sets limits on the physical laws and properties of our universe, cosmologists in the 1960s began to formulate the idea that *our simply being here* must limit the physical laws and properties of our universe to those laws and properties capable of generating carbon-based life forms like us. Furthermore, if some periods in the evolution of such a universe are consistent with life's existence and other periods are not, then our being here even sets limits on the stage of development that our universe is currently in.

This intriguing idea has become known as the "anthropic principle," of which both a strong and a weak version have been articulated. The weak version restricts itself to the universe we actually inhabit and reminds us that the nature of this universe and its period in history must be consistent with life's developing on earth. The stronger and more controversial version extends itself to consider other possible universes and concludes that only at certain times in certain kinds of universes (like ours, now) could life arise at all. For the purposes of our inquiry as to whether the laws and properties of the cosmos have anything to do with us, we need only consider the weak version, which is now widely accepted by cosmologists (whether or not they also accept the stronger version).

A good example of the anthropic principle in action is the question of whether the universe is currently expanding or contracting. The traditional view of the relationship of us and our thoughts about it to this "empirical" matter is that the current state of change in the size of the universe has *nothing to do* with us and our thoughts about it. It is strictly an a posteriori matter for scientific discovery.

The anthropic principle calls this traditional view into question. Stephen Hawking has elucidated this example of the weak version in his brilliant *A Brief History of Time* (1988). In his wide-ranging considerations about the limiting conditions of a physical universe that has manifestly given rise to us as living beings, Hawking takes up this question about whether the universe is currently expanding or contracting. He explains that the state of matter and energy we could expect to find when our universe is expanding is quite different from the state of matter and energy we would find when it is contracting. At times when the universe is contracting, all the matter and energy in it are almost completely disordered, with all stars burned out and all protons and neutrons having decayed into light particles and radiation.

Hawking suggests that during such contracting periods in the history of the universe, there could not exist the complexity of ordered matter that would be needed to give rise to any living organisms—including organisms capable of asking such questions as whether the universe is expanding or contracting. In our particular case, we know that the development of living organisms required such intricate molecules as RNA and DNA to evolve before the question could be asked, but we can assume that some complexity of matter would be needed to generate any form of life capable of asking such a question. Such complexity only exists during expanding phases of the universe, according to Hawking, and only during certain epochs in the expanding phase at that. If this is true, then any time the question arises about the current change in the size of the universe, the answer will *necessarily* be that the universe is expanding, since the expanding size of the universe is an a priori condition for the existence of such a thought (and not an a posteriori discovery *by* thought, as the traditional scientific view imagines).

This is not to say that our thoughts are *causing* the universe to expand. The argument is, rather, a theoretical physicist's version of *deduction*, and it follows in a wonderful tradition in the philosophy of knowledge. Descartes, starting with his own thoughts, was able to deduce only that he existed. Kant, attending to more subtle features of his own thoughts, deduced that if his thoughts exist, then so must the objects of his experience (that is, the rest of the physical world). The anthropic principle now extends the deduction: given that my thoughts exist, I can deduce not just that the world exists, but I can deduce quite a number of the properties of that world. That its size is expanding is one such property, but Hawking also explains how the same is true for many of the physical constants and laws of thermodynamics measured and calculated by physicists: if those constants and laws were other than what we find them to be, we would not be around to measure or calculate anything. Given a subtle enough understanding of the structure of the universe, the values of those constants and the shape of those thermodynamic laws may be *deduced* from the fact that "I think" in an intriguing extension of the way Descartes deduced his own existence and Kant deduced the existence of objects external to himself. With apologies to Descartes, we might say, "I think, therefore the universe is expanding."

There is something intriguingly anti-Copernican about the anthropic principle. As the cosmologists John D. Barrow and Frank J. Tipler explain in their definitive *The Anthropic Cosmological Principle* (1986), Copernicus expelled human beings from the center of nature and insisted that any scientific investigation forevermore begin by denying that we occupy a privileged position in the universe. But in one sense we human beings *do* occupy a privileged position in the universe, in that for most of its history the universe was not (and will not be) observable at all, because no life can exist to observe it. Only a small window of time is available for "knowing" the cosmos, and we are in this sense most privileged to be the "knowers." Indeed, when cosmologists apply their faculty of reason to understand the universe around us, they observe subtle inconsistencies and inadequacies in simpler conceptions of the cosmos that grapple their understanding of it to include knowledge of even those periods in history when no one is around to observe anything.

While it is not at first obvious that our living existence is a larger context that subsumes what laws of nature physicists can know, the anthropic principle demonstrates how this relationship holds. The realm of possible cosmological laws physicists can know is constrained by our being here, not because human thoughts cause the universe to be the way it is, but because there are constraints on the kind of universe that could cause physicists to come into existence. (This is perhaps more reminiscent of the theological argument that says God could not exist without us than of the argument that we could not exist without God!)

While expositions of the anthropic principle quickly become tied up in the abstruse jargon of theoretical physics, we might think of it more in terms of what Hegel told us in different language: that reality itself exists in a dialectical relationship with us. Indeed, in our discussion of the construction of experience in Chapter 3, we saw that Kant's critique of Hume insisted that "the understanding does not derive its laws from, but prescribes them to, nature." We also saw that Kant added that this "seems at first strange, but is not the less certain." Now we see that most of the theoretical physicists of our time agree with him!

Whether we are talking about Kant's understanding of the a priori preconditions of any possible human experience or about the implications of the anthropic principle, we are certainly ready to challenge that hard-nosed natural scientist who would take the "Earth-Mother Goddess" view of Mt. Everest to be merely a romantic notion and still maintain the belief that the physical nature of the world has "nothing to do with us." Indeed, we might want to think instead about a continuum of the ways and degrees to which we humans and our thoughts relate to the world around us: with the tables and chairs we actually *create* at one end, and with the basic laws of the working of the universe at the other end, not created by our thoughts but still constrained by our being here because of deeper connections that relate both to them and to us.

We shall return to the anthropic principle in later chapters as one of the perspectives from which we can reconcile the two apparently contradictory views of our basic concepts and of these basic laws of nature: the one that sees them both as necessary a priori preconditions of any possible experience, and the other that sees them both as contingent products of evolutionary experimentation. What is clear already is that we have moved into a much more fluid view of the philosophy of knowledge. Hegel had a wonderful expression for this fluidity in his own description of *The Phenomenology of Spirit* when he wrote that his book "deals with the *becoming of knowledge.*" Through these many subtle relationships between thoughts and things, our knowledge is not merely an activity and not even merely an achievement but is an ever-evolving adventure that *becomes* over time.

In his superb elaboration of Hegel's thought, the contemporary philosopher Robert Solomon captures this spirit in his idea that Hegel, in a rather naturalistic way, subsumed "knowing" under a larger context, namely *living.* Our ideas are thus not merely right or wrong, but are products of the times; and even more than that, our ideas are ways of *dealing with* the times, ways of *accomplishing* something. Even in the field of theoretical physics, we have just seen how knowing about the laws of the universe has been subsumed under the larger context of our living existence. The reason our original distinction between thoughts and things so grossly oversimplified the dynamic process of knowing is that it was completely taken out of the context of living. Ultimately, knowledge itself does not function as an autonomous system because, as the existential philosopher Martin Heidegger also made clear, *knowing how* (to live, to satisfy desires, and so on) is as genuinely a part of knowledge as *knowing that* (there is a table in front of me, the chair has four legs, and so forth).

Once knowing gets subsumed under this larger context of living, we are freed to consider other features of human experience and never really leave the philosophy of knowledge. As described in Chapter 1, an exploration within the contexts of remarkably different disciplines of the relative contributions of ourselves and of the world around us ultimately reveals the deeper, close relationships linking those very disciplines. While we will now turn our primary attention away from the traditional field of philosophy, we have already seen how other human activities—from the experience of discomfort in the nursery to the economic activity of manufacturing and selling tables and chairs to the theoretical discoveries of cosmologists—are inseparable from any discussion of a complete theory of knowledge. That more complete theory demands, however, that we turn our full attention to some of these other contexts so that they can be developed before we return to support that complete theory of "knowing as living" in a more satisfying way.

9

The Facts of Life

In the preceding chapters we have begun to see how the work of empirical psychologists such as Piaget, and even theoretical physicists such as Hawking, might help us uncover subtle nature-nurture contributions to our knowledge of the world. By thus introducing the physical world as biology conceives it into our investigations, we must accept the existence of intimate connections between even the most abstract philosophical questions and empirical studies in natural science areas such as biology and physics, and indeed in the social sciences (such as anthropology) as well. The question of which of the natural and social sciences are most relevant is itself a scholarly question of significant import.

If we begin with the natural sciences, biology is clearly our most relevant starting point. As mentioned in Chapter 1, advances in biology have long been known to fuel swings in the nature-nurture pendulum. More specifically, we might want to focus on neurobiology, the burgeoning field which promises to be especially significant as it attempts to uncover the actual biological bases of cognition, emotion, and other aspects of conscious and unconscious human experience. Philosophers and psychologists have long recognized that advances in our understanding of the brain might someday set limits on acceptable theories of the mind. Certainly as we understand ever more about the functioning of the human brain, philosophical or psychological theories about our knowledge or our ethics must fit comfortably with the actual neural mechanisms that give rise to beliefs and values. It is premature to demand a perfect fit, since all of these disciplines are dynamic and always evolving into more mature understandings of their fields of inquiry. But a reasonable fit is already demanded, given the explosion of new advances in neurobiology over the recent decades.

By focusing on the biology of the human brain over the next few chapters, we shall achieve a fuller view of some of the philosophical questions already discussed and also set ourselves up for explorations of other related fields still to be explored. In our twin experiment, for example, we took some pains to try to distinguish between two evolutionary strategies. In what we called the "nature approach," our genes would dictate brain structures that determine (establish the "grid" for) basic concepts found to be advantageous over earlier periods of evolutionary history. In what we called the "nurture approach," our genes were set up more to enhance the brain's plasticity to mold itself to whatever postnatal environment it might confront. That entire discussion becomes almost metaphoric when we now focus our attention on genes as actual biological entities and review some of the "facts of life" as understood by neurobiologists.

If we simply look at the relevant numbers from genetics and neuroscience, we find that there are approximately 10 billion (10^{10}) to 100 billion (10^{11}) nerve cells in the human brain, which establish somewhere between 100 trillion (10^{14}) and 1 quadrillion (10^{15}) cell-cell connections. In contrast, the total number of human genes is currently estimated at between 50,000 and 100,000. The sheer difference in orders of magnitude between the number of cell-cell connections established and the number of genes responsible for them makes it obvious that nature had to opt for some strategy that would enable events which follow conception to contribute to the shape of the human brain.

The facts of neurobiological life begin with a sperm and an egg forming a single cell, whose fixed genetic makeup must produce the billions of brain cells that will become responsible for all thoughtful interaction with the world. Since the sheer numbers involved suggest that the genes in this single fertilized egg cannot contain the entire score to orchestrate the trillions of cell-cell connections to be created and harmonized, nature has employed several ingenious strategies to solve this problem. To the extent that the genetic information in that fertilized egg might be called our "natural endowment," these strategies by which our genome enlists the help of the environment to contribute to the shape of our brain might be viewed as "nature harnessing nurture" for its own purposes. Although the detailed process by which this one cell becomes a human being with a human brain is highly complex and still not well understood, we can already identify a number of strategies by which nature has harnessed nurture, and the rest of this chapter will be dedicated to a brief description of some of the most important of them.

The development of billions of nerve cells from a single fertilized egg during embryogenesis requires two different steps. The first step is called *determination*—the process by which certain cells become destined to become *nerve* cells, in contrast to, say, muscle cells. Determination is how the fate of the cell becomes fixed. The second step is called *differentiation*—the process by which a given cell, having been determined to be a nerve cell, is established within a particular neural network with particular connections within the brain.

The cell's own genome obviously plays a central role in determination and differentiation: either can be changed by a single mutation. But even though the details of these processes are still poorly understood, enough is known about both to be sure that neither is entirely genetically specified. Whether a given cell will end up as a nerve cell and, if so, whether it will play a role in vision, memory, language, or emotion are questions whose answers depend upon what is going on *outside* the cell as well as inside the cell. While we have thus far talked about the "environment" as that which is outside the organism, we must now also remember that in addition to considerations of this external environment, the formation of a given nerve cell and its connections might also depend upon the *internal environment:* the local physical and chemical milieu offered by neighboring cells within the brain itself. This "microenvironment" now implicates the genome of other neighboring cells (which divided from the same egg—a nurturing by one's own nature!).

A famous example is the development within the embryo of the cells that will eventually become our brain and nerves (the neuro-ectoderm). It has been known for over fifty years that a given cell becomes part of the neuroectoderm in part because of its position in the embryo next to cells that are developing into muscle cells (that is, the mesoderm). In Hans Spemann's famous experiments in the 1930s, not only did removal of the mesoderm prevent the develop-ment of the neuroectoderm, but transplantation of mesoderm next to cells that would have become skin cells caused some of these to become nerve cells!

Even more dramatic examples of the effects of local environment can be seen in the further maturational steps through which embry-onic nerve cells (having been determined and differentiated) weave themselves together into the intricate pattern we call the brain. All steps of this maturational process—from neuron growth and elon-gation to migration into position within the brain, to the formation of specific synaptic connections—are subject to the influence of the local environment. Of course, these local influences have limitations. Just as a cell that has now become a nerve cell can no longer switch and become a muscle cell (even though an earlier cell might have become either), there are similar critical periods of development after which the migration within the brain and elongation of most neurons appears to be finished.

At the end of this maturational process, these internal environ-ments have interacted with the genetic information in that fertilized egg, so that the remarkable outcome produces human brains that look rather similar in their gross appearance but which, like snow-flakes, are different at the fine, cellular anatomical level in every two human beings who have ever lived or ever will. Even identical twins, who share the same genes and the same uterine environment, have brains with different wiring diagrams. This is in part because the true *micro*environment within which each of their individual neu-rons is maturing is different enough to produce different effects, and also because of a certain degree of randomness in cell movements and chance molecular events that occur within cells during development. (This developmental "noise" probably accounts for the fact that identical twins have different fingerprints as well.)

Some of the most influential environmental contributions to the developing shape of the brain—through the processes of determination, differentiation, and further maturation—occur when the environment is still the womb of the mother. In the nature-nurture debate people often forget that the day of birth follows many months of environmental exposure. If a pregnant woman takes certain medications or is exposed to x-rays at critical periods of the pregnancy, the resultant congenital malformations in the infant (a cleft palate or a missing limb, for example) can hardly be said to be the results of nature and not nurture. The influence of the local hormonal environment is particularly striking. In animal experiments, the administration of testosterone to female fetuses in utero gives rise to hermaphroditic offspring that exhibit male patterns of sexual behavior. While part of the mother's "internal" physiological nature, the hormones that circulate from her own endocrine glands (ovaries, thyroid, adrenals, and even the placenta itself) nurture the shape of the developing brain as a part of the brain's local environment.

These dramatic malformations and sexual changes are obvious examples, but we still know little about the more subtle in utero environmental effects of a minor "cold" virus during the third month, exposure to loud music in the sixth month, or changes in the hormonal environment during the eighth month due to anxiety from the mother's taking a leave from her job.

These examples remind us how important it is to define *congenital* in terms of what exits at birth but is *not* a matter of heredity. Recall, for example, the debate over whether basic concepts such as object or number are "congenital" or "learned." We can now see that even Leibniz's nature-side argument for the possibility of "dormant congenital concepts" could not be won by proving that we are born with a program for these dormant concepts, since the environment may well have contributed prenatally to the shape of even those programs we are born with.

Is nature taking much of a risk in harnessing the nurturing of the uterine environment to help orchestrate the determination, differentiation, and further maturation of our developing brain? While we often focus on the womb-to-womb differences between these environments (because of exposure to x-rays, drugs, or hormonal shifts), the next fact of life we have to recognize is that human intrauterine environments have much more in common with one another than they have differences. In fact, when a given uterine environment is significantly different from nature's prescription, fetal development does not proceed. This is no small point. To the extent that our discussions so far have focused on the thoughts and feelings and values of actual human beings, our biologist's view of the matter gives us a very different picture of what Kant called the "necessary conditions for any possible experience." Those conditions now include a successful union of sperm and egg with more or less the correct genetic material to combine and reproduce, and a good enough uterine environment to be harnessed by nature throughout the nurturing of the developing fetus.

This dynamic is no different once the baby has been born. We saw above how normal human development depends on the nurturing of some maternal environment, so that babies who do not remain "attached" to others (after the physical, umbilical attachment has been severed) not only fail to have knowledge and make moral distinctions, they fail to survive. If our first of nature's strategies for harnessing nurture is to enlist the support of a uterine environment with fairly high specifications for the job, then the second strategy is to enlist the support of adult members of the species which human infants require for survival during their prolonged period of dependence.

Homo sapiens has a longer period of dependency during early life than any other known species. Put in the context of the species' numerical problem of too few genes to specify so many cell-cell connections in the brain, this prolonged dependency may also be understood as a way to harness nurture in order to help orchestrate the development of a brain capable of having knowledge of the world outside itself. To enhance this strategy, nature, as we have seen, has supplied directly a number of reflexes to harness the influence of nurture more effectively. Although Piaget looked upon simple motor reflexes such as sucking and grasping as the beginnings of "intelligence" through which the infant begins its own construction of a suckable, graspable world, further behavioral and neuroscientific research has shown that the infant's motor reflexes (a presumably innate, genetic ingredient) are not so simple, and can in fact be quite complex.

Over the next two chapters we will focus on the development of the visual system of the human brain, and so we might take here as an example of adaptive reflexes some of those that control the infant's eye movements. In regulating the behavior of the muscles that move the eyes, these reflexes channel, restrict, and to some degree help "program" those encounters with the visual environment that will be available to contribute to the shaping of the visual system made possible by the brain's plasticity.

The best example of this is the infant's reflex to turn its eyes toward the caregiver's face. This reflex is very adaptive, in that it is one example of an infant's natural beguiling which helps the infant secure the life-sustaining attachment of the mother, as we saw in Chapter 7. The mother's sense that the new baby "seemed to know me from the start" is in part secured by a motor reflex which makes babies focus on certain stimuli. Studies have shown that the stimulus that orients the baby's eyes in a particular direction is not actually the mother's face at all but certain specific features of high contrast that may be found in a pair of eyes and also in most toys created for infants (thanks to the discovery of this same reflex by toy developers and market researchers).

Three points can be made about a reflex such as this. First, as mentioned, it is probably part of how the baby secures the attachment of caregivers—reinforcing other reflexes such as the grasping of the mother's garment and the innate sucking at her breast—and so helps to harness a protector through the coming long years of dependency for survival. Second, this same motor reflex dramatically increases the probability of an infant's receiving sensory input from a particular aspect of the environment, namely, human faces.

What Will Babies Smile at More?

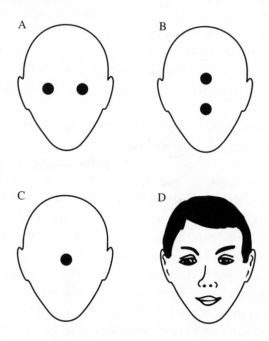

These patterns have been used to study the stimuli that will elicit a smile in babies. For babies six weeks of age, patterns A and B are more effective than C and D, the critical features appearing to be multiple spots of high contrast. As babies mature, the face image progressively becomes more effective in eliciting a smile than any of the dot patterns.

Given the importance of the interpersonal environment to successful adaptation throughout life, this motor reflex serves the infant well by assisting those parts of the brain that need to glean crucial yet subtle information from human faces (such as the emotional state of the person being observed) in their calibration very early in life.

Third and finally, we can ask again whether nature is taking much of a risk in harnessing the input of faces in the environment to nurture the development of those parts of the brain that will need to react to the emotional states of others and relate to others in a great variety of subtle ways. The answer is that, although we tend to focus on the differences between nurturing environments, once again there are probably even more important similarities among any environments that can sustain human life than there are differences. Rather than trying to evolve a large number of genes to code for a preformed understanding of facial expressions in relation to the intentions and emotional states of others, nature is able to concoct a single reflex to orient eye muscles to areas of high contrast (such as eyes) and let the environment do the rest of the job, since any human being that will survive to an age that requires this understanding of facial expressions will have had to have someone (or likely more than one person) involved in his or her nurturing in order to survive. Although Darwin was impressed by the apparent genetic basis of the muscle programs involved in the facial expression of emotions (finding similar coordinated facial muscle movements for fear, anger, happiness, and so on across many species), he could not have imagined the many subtle mechanisms harnessed by nature to enable the brain of another animal *observing* these expressions to pick up their meaning and implications during postnatal development.

Over the next two chapters, we shall explore in more detail the way the brain interacts with its environment in order to secure the nurturing of certain environmental features that are in some cases constant enough across the entire species to support even our basic human concepts such as object, space, causation, and so forth. This plasticity of the nervous system to shape itself to these constancies of the world may therefore be understood as a further strategy employed by nature to harness a nurturing environment in orchestrating trillions of brain cell connections with comparatively few genes specified for this purpose.

For now, however, we might just place this effort in the context of all the preceding chapters, where the Hegelian challenge to study the "contributions of the world of things to the structure of our thoughts about it" might be reframed in our biological idiom as the "contributions of the environment to the structure of our central nervous system." We have seen how the central nervous system must inevitably depend upon the environment to participate in its developing structure. This is why a biologically informed philosophy of knowledge must take seriously the question of how much constancy we can take for granted in the environment of different (and even, perhaps, all) individuals. For now, it is only worth noting the obvious: that the environment as biology conceives it is *everywhere* and *unavoidable*. In the spirit of the anthropic principle, we take the environment to be constrained to have certain features that are necessary for the sustaining of any life forms of the sort that might ask questions about nature and nurture. It may well be that certain broad Kantian categories, such as object, causation, space, and time, are just a few examples of the necessary features of any such nurturing environment.

"Necessary" here is again no longer taken in the strict philosophical sense, but in the biological sense that without these environmental contributions development would not proceed in a way that would yield organisms capable of asking such questions. In fact, even less basic (even more a posteriori) ideas can probably be depended upon across environments in this way. When children come to have experience of feathers, wings, and fur, they come to associate wings with feathers and not with fur. When we look across cultures over our entire planet and notice that all children make this same connection in this way, we need not postulate some gene that directs a brain structure that makes this association. If, in our everywhere and unavoidable biological world, there are environmental features supporting the association of wings more with feathers than with fur, then this ubiquitous association in each individual mind can indeed be said to arise from nature—but only insofar as nature has been crafty enough to harness nurture in directing our understanding of the world.

10

The Perceptive Brain

Each of our divide-and-conquer strategists offered a picture of the human mind that separated the mind into a number of identifiable parts upon which the influences of nature and nurture could act. Our expanding vision of the nature-nurture debate has now shown us the relevance of the actual empirical world to our studies of the mind and focused our attention on the human brain as the organ of greatest relevance. It is therefore only fitting that neurobiologists who study the mind from this perspective also divide it into a number of parts with identifiable functions. In the illustrations on pages 23 and 38 we saw the structures of the mind as envisaged by Freud, Plato, and Kant. In the same spirit, the illustration on page 104 is a crude diagram of the left side of the human brain showing some of its most important surface features. This view of the surface structure of the brain is not entirely dissimilar from Kant's model, since we find discrete sensory input areas (which together might be thought of as our faculty of sensibility) and a large central processing area traditionally known as the *association cortex* (which might be thought of as our faculty of understanding). In fact, the main difference in this model is that, in moving on to the brain, we now remember that we humans have, in addition to our sensory input systems and our conceptual grids, a set of what might be called output systems. These are the *motor systems* that we use to move about the

The Human Brain

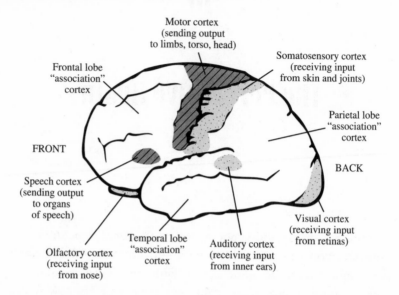

The left side of a human brain, demonstrating how the cortex may be divided
into three separate "faculties." Input processing systems (for analysis of incoming
sensory data) are dotted, output processing systems (for movement and speech)
are striped, and central processing systems (the "association" cortex of each lobe)
are shown in white.

world and to speak (and to direct the rest of the body's own glandu-
lar systems to release various hormones). We have actually already
seen the complex interactions among these three brain areas in the
last chapter, as when a motor reflex directs our eyes to receive higher
densities of certain aspects of the visual environment which might
lead to different associations in the conceptual areas of the brain
(our motor system influencing our sensory input system, leading to
changes in our conceptual system). It was no accident that Piaget
called the first step we take in our sophisticated understanding of the
world "sensorimotor intelligence"!

To those unfamiliar with neuroanatomy, it may seem strange to
relate the functioning of the brain to the surface features in this

model, but the reason for this is quite simple. The brain evolved from the *inside out*. That is, at the center of the brain sits the brain stem, which connects the brain to the spinal cord. The brain stem is concerned mainly with *autonomic* functions: maintenance of blood pressure, fluid balance, breathing, wakefulness, and so on.

Surrounding and enveloping the brain stem sits what is loosely termed the *limbic system*. The limbic system is concerned with such things as emotion and memory. This developed slightly later in evolution than the brain stem (limbic system structures are present in all vertebrates). As we move from the inside outward, we find that the brain stem plus the limbic system in humans thus together looks remarkably like the brain of a reptile.

The third layer, latest in evolutionary development, surrounds and envelops the brain stem and limbic system; it is called the *cerebral cortex*. A cerebral cortex becomes increasingly identifiable in animals as we move up the evolutionary tree—it is highly elaborate in all mammals—and its size and complexity also increase as we move within the mammalian class from rat cortex to monkey cortex to human cortex. Physically, the human cerebral cortex is only two millimeters thick, but its thickness belies its complexity. Because of its convoluted topography, the cortex has a surface area, when spread out, of about 2,000 square centimeters. Under every square millimeter there are some 100,000 nerve cells, making for a staggering total of some 10 billion neurons in the cortex alone, with cortical cell-cell connections numbering thousands of times greater than this.

There are, of course, many other important structures in the human brain, such as the cerebellum at the back (responsible for, among other things, coordination) and the so-called basal ganglia between the limbic system and the cortex (responsible for, among other things, organizing volitional movement). Although some of these other brain structures play very important roles in human experience (especially those deeper limbic structures, as we shall see in Chapter 12), this chapter's brief review of human brain structure will focus on the cortex.

Three types of evidence support the division of cortical brain regions into sensory input areas (dotted), motor and speech output areas (striped), and "association" areas (white): anatomical evidence, functional evidence, and neurochemical evidence. I shall save the neurochemical evidence until the next chapter, when we focus more on brain mechanisms of plasticity. Let us begin here with the anatomical evidence.

Perhaps the easiest way to understand this division of cortical areas is in terms of the *microscopic* anatomy of the brain. If you pluck off a bit of the cortex and stain it to look at its cells under the microscope, you find that the two-millimeter thickness can be divided into six layers of cells over most of the surface of the brain. The relative proportions of large versus small cells in these six layers reveal three distinctive patterns that characterize different regions of the cortex. All of the sensory input areas of the brain have *small* cells in all six layers (so-called granular cortex), all of the motor and speech output areas have *large* cells in all six layers (so-called agranular cortex), and the remaining association cortex has various alternating layers of large and small cells (so-called homotypical cortex).

Although this point of microscopic anatomy is most obvious to neurobiologists, much more intuitive anatomical evidence for our map of brain regions comes from the connections of our sensory and motor organs to these various areas of the brain. The eyes, for example, are connected fairly directly to the visual sensory input area at the back of the brain, our ears are connected fairly directly to the auditory input area, our skin to the sensory area for touch, our mouth to the speech output area, our arms and legs to the movement area, and so forth.

More interesting evidence for our cortical map comes not from anatomical but from functional considerations. The most direct way to investigate the function of different brain regions would perhaps be to open up the skulls of some volunteers and stimulate various parts of their brains, asking them what they are experiencing with each stimulation. Although it might appear that ethical considerations would prohibit this kind of direct investigation, neurosurgeons have occasionally performed the procedure when operating to remove a tumor or cyst from the brain. Having encased the brain safely within the skull, evolution did not waste the energy of putting pain sensors inside the brain, so neurosurgery can be done safely and painlessly on an awake, alert patient after a local anesthetic is used to allow for the creation of a skull flap to expose that part of the brain needing repair. This can become preferable to a general anesthetic in certain cases when the cyst or tumor has grown near crucial sensory, language, memory, or motor areas of the brain: the surgeon can use these electrical brain stimulations and the patient's verbal reports to map out important areas that must be left intact to avoid unnecessary blindness, paralysis, loss of language or memory, and the like.

In pioneering this technique, the great neurosurgeon Wilder Penfield provided the first direct evidence for our map of cortical brain areas. In now-famous studies Penfield found, for example, that stimulation of the primary visual area of a person's brain produces the experience of very simple hallucinations—points or lines of flickering light, whorls of color. In contrast, stimulation of the cortical association regions near the visual area produces the experience of complex hallucinations—including completely formed visual images from long-"forgotten" memories! There appears to be a shift in function from the primary visual cortex's representations of the basic perceptual building blocks of visual experience (lines, colors) to the central processing area's capacity for integrated, complex visual experiences. We may take this as direct support for a map of the brain in which crude sensory information is received and assembled in one area (sensibility) and then integrated and synthesized in another (understanding).

Perhaps the most compelling evidence for a similarity between the anatomical structure of the brain and Kant's notion of a faculty of sensibility separate from a faculty of understanding comes from the discovery of receptive fields. A receptive field in visual processing is the area of the visual environment that provides input to a given area of cells. If we randomly pick a square of cortical cells of size one millimeter by one millimeter and ask "How much of the total visual environment is providing input to these cells?" the answer will depend very much on where we pick the square. In the primary visual sensory area, for example, that square of cells would be receiving information about what is happening in a very small sector of the visual field: on the order of one degree of arc in diameter. This small receptive field will have increased by two orders of magnitude by the time we move to the border between the visual sensory area and the association cortex, where the same size square of cells would be receiving information from more than half the visual field: on the order of 100 degrees of arc. We may take this as strong evidence that more and more *integration* of information occurs as the sensory information is processed by the complex brain mechanisms in our association cortex.

The effects of certain injuries and diseases on experience are also illuminating. People who lose their visual cortex (say, because of a tumor) go almost totally blind. But if they develop a tumor or stroke in the border region between the visual cortex and the association cortex, they manifest an unusual condition called visual agnosia. In this condition, described vividly by Oliver Sacks, people can correctly identify the orientation of E's on an eye chart but have lost the ability to identify, name, or match even simple objects in any part of their visual field. They appear cognitively, not visually, impaired. In essence, they can *look* (they still have 20/20 vision or whatever acuity existed before the tumor or stroke), but they cannot *see* (that is, make sense of what they are looking at). We might think of this in Kantian terms as a disconnection between sensibility and understanding—the faculty of sensibility can analyze its sensory input, but the *meaning* of the information is lost without the synthetic work of the faculty of understanding from which sensibility has become disconnected. (In considering the potential disconnection of sensibility from understanding, Kant told us that "thoughts without content are empty, intuitions without concepts are blind." The latter state might more accurately be called not blindness but visual agnosia.)

Having divided the brain into input, central processing, and output systems according to anatomical and functional evidence, one can look comparatively at how much of the brain has been allocated to each area in various animals. What we find is that humans have the largest association area as compared with input plus output areas, and in fact the ratio of association to input plus output areas gets smaller as we move from humans to monkeys to rats. Philosophers and biologists alike speculate that complex cognition and even consciousness are not all-or-none phenomena, and that higher mammals and certainly higher primates must have some rudimentary capacity for what we have called "thinking about things." On our analogy of the association cortex with the faculty of understanding (Kant's *sine qua non* of thinking about things) this evolutionary ratio gives some quantitative sense to these speculations. Placed in this context, it is no coincidence that the human brain has the largest ratio of association to input plus output areas, since we presumably spend that much more time thinking about things!

When we were peering at plucked-off bits of cortex to characterize those quantitatively distinct regional patterns of large and small cells among the six stained layers under our microscope, we might have also turned the magnification down and noticed an interesting *qualitative* difference in the pattern of how generally organized or disorganized the cellular connections look in different areas of the brain. At this gross level where larger bundles of cells can be observed, the neuroanatomical pathways that connect the sense organs and muscles to their relevant primary sensory and motor cortical areas (and also to some extent the connections within those primary sensory and motor cortical areas) appear to be much more hardwired and more neatly organized than the anatomy in the association areas of the cortex (where the cellular connections appear to go every which way in more of a tangle). This is part of the reason that neurobiologists have been more successful in discovering the detailed anatomy of the input and output areas of the brain (Kant's faculty of sensibility and our newly described "output" faculties) than they have to date in the association areas, but advances are now also being made in the anatomy of what Kant as a neurobiologist might have labeled our faculty of understanding. Let us look at the structure of the visual system as one example.

The neural pathways that connect each eye in the front of the head to the primary visual cortex at the back of the brain are highly organized bundles of nerve fibers that initially exit the back of the eye as the "optic nerve," made up of about a million nerve fibers. In their organization, these fibers preserve the relative positions of the cells on the retina of the eye to which they are connected. The retinal cells, of course, are responsible for receiving visual stimuli.

Shortly after leaving the eye, the optic nerve bundle does what most sensory and motor nerve cells do in our bi-hemispheric brain: it reorganizes itself to send information from the left half of experience to the right side of the brain and information from the right half of experience to the left side of the brain. While a sensory modality like touch can accomplish this just by crossing all the fibers coming from each arm and leg to the opposite side of the brain, the visual system is more complicated. Only half of the fibers from each eye cross over, so that what ends up on the left side of the brain is not all the right eye fibers but the fibers from *each* eye containing information from the *right side of the visual field* (which includes half of the fibers from each eye, since both eyes receive information from both sides of the visual world). Similarly, all visual input from the left half of the visual field gets sent from both eyes to the right side of the brain. The evolutionary reasons for this cross-over and for the separation of hemispheric functions—for example, the tendency for language to be processed in the left hemisphere—are still not known.

With only one cell-cell connection ("synapse") from the eye in front to the visual cortex in back, the first neural steps in my seeing those trees in my yard follow highly stereotyped cellular pathways. Within the primary visual cortex, the information about those trees is processed through a series of three or four levels of cell-cell connections, which we shall discuss shortly.

As we continue to follow the flow of visual information out of the visual cortex and into the visual association areas, things start to get more complicated. Exiting from the visual cortex on each side of the brain is not one but multiple parallel tracks leading into different sections of the association cortex. These distinct multicell cortex-to-cortex connections appear to be responsible for different aspects of visual experience. There is, for example, compelling evidence that one pathway leading down into the temporal lobe processes "object vision" (enabling me to see the shapes and colors of those trees as objects), while a separate pathway leading up into the parietal lobe appears to process "spatial vision" (enabling me to perceive the spatial relations among the trees in my garden, but not their intrinsic qualities, which are processed by the other pathway). With this same bifurcation of dozens of other similar pathways coming from each of the sensory input areas, the complex, tangled appearance of the cells in the association cortex is hardly surprising!

Clinical neurologists have for some time known that the "spatiality" of human experience is synthesized in the inferior parietal lobe (the ideal spot, situated as it is between inputs about how the world feels, how the world looks, and how the world sounds), since a tumor or stroke in this area of the brain leads to dramatic impairment of spatial perception. Thus, such patients have normal visual acuity (they can look) and can identify, match, and name objects (they can see), but they cannot locate objects in space. The most dramatic examples of this are patients with the syndrome known as *hemispatial neglect,* who have lost the inferior parietal lobe in their right hemisphere and are completely unaware of the left side of their world— including their own left arm!

With these discoveries of identifiable associative cortical regions for the processing of the spatiality and object-ness of our experience of the world, it is important not to become overly optimistic that the entire association cortex will unfold like Kant's grid of categories which, he argued, underlie all human experience. Indeed, if anything, biological research reminds us how such distinctions as those between sensory input systems and central processing systems are very ill-defined, with no clear boundary between the two. Precisely where we draw the line between the highest orders of input processing by our "faculty of sensibility" and the first stages of the conceptual work of our "faculty of understanding" is somewhat arbitrary. Indeed, here in the biological idiom we begin to see how the boundary between our faculties becomes blurred even as the existence of some distinction is supported—a theme to which we shall return in Chapter 14.

Neurobiologists have also made great strides in understanding the

The Spatiality of Experience

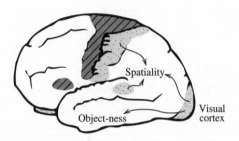

The figure above shows how multiple parallel pathways leave the visual cortex to construct different features of experience in different association areas. Although only the pathways leading to the construction of our experience of spatiality (in the parietal lobe) and object-ness (in the temporal lobe) are shown, over two dozen other separate parallel streams of visual information have been identified. The area for spatiality also shows how input is simultaneously arriving there not only about how the world looks (from the visual cortex) but also about how the world feels (from the somatosensory cortex) and how the world sounds (from the auditory cortex).

When a tumor or stroke damages the area marked "spatiality" on the right side of the brain, the dramatic result, called hemispatial neglect, can leave patients with no experience of the left side of the world – including their own left arm! Shown on the right side of the figure is the outcome when such a patient was instructed to draw a clock with all the numbers and a daisy in a pot.

details of the architecture of the visual cortex itself, which has now been divided into a large number of subsections. The first of these is called the *primary visual cortex*. It receives information coming from the retina and processes that incoming information through those three or four levels of cell-cell connections. The primary visual cortex then sends this processed information on to the *secondary* and *tertiary* visual areas. There it undergoes further synthesis with information coming from other parts of the visual field.

The primary visual cortex in the back of the brain is the first point in the visual system at which fibers carrying information from the two eyes converge upon single cells. Although the nerve bundles carrying fibers from the eyes to the back of the brain organize themselves shortly after leaving the retina so that half of the information from each eye combines on each side of the brain (allowing each side to process information from the opposite side of the visual world), these left and right eye fibers are still just parallel streams of information; although they continue to preserve the relative layout of the two retinas, they do not *connect* with each other until they reach the very back of the brain. It is therefore not surprising that scientists have focused large amounts of effort on the primary visual cortex in their attempts to dissect the functional architecture of the visual input system. The results of this work have been extraordinary. Within the primary visual cortex scientists have discovered a highly complex anatomical organization wherein columns of cells are organized to react to specific types of visual input coming from the retinas.

The discovery of this columnar architecture of the primary visual cortex as well as other primary sensory cortexes (for touch, hearing, and so on) has opened up many new areas of exciting research, some of which will be discussed in Chapter 11. As an example, specific columns of cells have been discovered in the primary visual cortex for angular orientation, spatial frequency, direction of movement, color, depth sensitivity, and other such perceptual building blocks out of which the brain reconstructs the visual environment. From out of these crude bits of sense data gleaned from the information received by the retina and sent on to the brain, our visual system constructs our visual experience of the world.

In the spirit of Piaget, we can immediately appreciate the similarity between the way we assimilate the environment in eating food and the way we assimilate the environment through our sensory perceptions of it. The complicated ways in which the retinal information is sent through a series of way-stations in the primary visual cortex to be sliced and diced according to highly specific perceptual properties is analogous with the way he described the assimilation of a piece of food through cutting, chewing, and digesting it and thereby making it a part of ourselves. As I sit writing this and have the visual experience of those trees in my yard, my central nervous system has likewise taken those lightwaves and forms and cut, chewed, and digested them to make those trees a part of my visual experience.

The view of the structure of the human brain that has been painted briefly in this chapter has thus focused on what Piaget called the assimilation of the environment by the organism. In the tradition of Kant, we have focused on the contributions of the structure of our organ of thought to any experience we can have of the world. You will recall, however, that assimilation is only half of Piaget's story of how we adapt to the environment. Let us therefore turn in the next chapter to the neuroscientist's version of how we *accommodate ourselves to that environment* and change the very shape of those neural structures in the process of that adaptation to the environment that we call "thinking about things."

As the founder of modern evolutionary theory, the English naturalist **Charles Darwin** (1809–1882) changed forever not only biological theories but theories of knowledge and values that hope to be grounded in the reality that the human brain is a product of evolution. Ever since Darwin, we have known that nature is "thoughtless," but thoughts themselves emerged from nature, as brains capable of thoughts arose through evolution. Darwin contrasted with the process he called natural selection the parallel process of "artificial selection," by which the thoughts people have about breeding animals or cultivating fragile crops can lead to "unnatural" outcomes for these naturally occurring things. He thus highlighted how such "artificial" thoughts are uniquely both a part of nature and apart from nature.

Often called the father of modern philosophy, the French mathematician, scientist, and philosopher **René Descartes** (1596–1650) developed a system of physics superseded only by Newton, and he invented the mathematical equations, exponents, and analytic geometric coordinates we still use today. Descartes's dualistic theory of knowledge highlighted the distinction between *things* that exist in the world and our *thoughts* about those things—thoughts that exist in the mind. Descartes challenged all future philosophers to find a way they could be certain that the thoughts in their minds accurately represent things in the world, and his view of the mind as something like a private thought-theater existing in relative isolation from the things in the world has permeated Western conceptions of the mind ever since.

The Scottish philosopher and historian **David Hume** (1711–1776) rose to Descartes's challenge. In exploring the ways things in the world write their qualities accurately into our thoughts about them, Hume (taking up the empiricist philosophy developed by John Locke) expounded the notion that the mind is like a blank slate, or *tabula rasa*, that receives these qualities from things in the world. In his moral philosophy, Hume developed the idea that most human values—such as justice—are artificial creations of the mind, devised to try to make the natural world turn out better that it otherwise would (a bit like Darwin's later idea of artificial selection).

The German philosopher **Immanuel Kant** (1724–1804) recognized that Hume's notion of things writing their qualities into our minds might make sense for such sensory qualities as color but not for such conceptual qualities as causation. Kant replaced Hume's blank-slate metaphor with a picture of the mind as composed of distinct parts or faculties, some of which merely perceive the sensory qualities of things, but others of which construct our experience of things by applying our mind's own concepts ("categories") to these incoming stimuli. In claiming that all the concepts we hold about things in the world actually come from our own mind—are an innate part of our nature—Kant thus asserted, for example, that Newton's laws of physics are not properties of things in the world at all but are properties we bring to our experience of those things. By making knowledge an activity and not merely a product of the mind, Kant, like Hume, also sought to rise to Descartes's challenge. The systematic theory of knowledge, ethics, and aesthetics that he developed has had tremendous influence on Western philosophy well into the twentieth century.

The German philosopher **Georg W. F. Hegel** (1770–1831) sought to make knowledge not merely a Cartesian product of the mind or a Kantian activity of the mind but a living achievement: an achievement arrived at by means of a dialectical process. In Hegel's scheme, our concepts are not an innate part of our nature but are ever-changing ideas that we can improve upon by constantly testing them against our evolving experience of the world. Hegel agreed with Kant that we construct our experience by applying concepts to that experience, but he disagreed about the source of those concepts, and in so doing issued a new challenge to the philosophy of knowledge: to find a way that the things in the world actually contribute to the shape of those concepts we apply in constructing our experience of them. Hegel sought to reunite thoughts and things, and his dialectical system—which emphasized not only the progress of concepts but the progress of history—has had a major impact on existentialism, Marxism, positivism, and modern analytical philosophy.

The English logician, philosopher, and Nobel laureate **Bertrand Russell** (1872–1970) is best known for his work in mathematical logic and for his advocacy of pacifism and nuclear disarmament (for which he was imprisoned in 1918 and 1961, respectively). Russell also raised the possibility that we can infer something about the world from the language in which it is (accurately) described. His ideas about the relationship between the structure of language and the structure of experience had a tremendous influence on his most famous student, Ludwig Wittgenstein (1889–1951). This relationship was later explored by Noam Chomsky, whose theory of an innate human grammar structuring any possible language would be, in Russell's view, a linguistic version of Kant's theory of an innate conceptual grid structuring any possible experience. Russell rebelled against Hegelian philosophy and essentially believed in the realist premise that the scientific view of the world is largely correct. Yet, he captured the essence of Hegel's dialectical progress of concepts in his description of Hegel as a "metaphysical comparative anatomist." Just as a comparative anatomist from a single bone sees what kind of animal the whole must have been, so Russell described Hegel as "seeking to deduce from any one incomplete piece of reality what the whole of reality must be, since each separate piece has metaphysical hooks which grapple it to the next until the whole universe is reconstructed."

Hilary Putnam, one of the most distinguished living philosophers, could also be called a metaphysical comparative anatomist—one who has found subtle connections in a naturalized theory of knowledge and values. These connections ultimately blur many old distinctions such as subjectivity versus objectivity, fact versus value, and synthetic versus analytic (what is or is not "true by definition"). Although Putnam maintains that philosophical questions cannot ultimately be reduced to arguments about evolutionary biology, his theory of knowledge and values is grounded very much in human experience and has been suitably dubbed "realism with a human face."

The Austrian neurologist and founder of psychoanalysis **Sigmund Freud** (1856–1939) applied his psychoanalytic insights not only to psychological but to mythological, anthropological, and religious phenomena. Like Kant before him, Freud divided the mind into a number of interacting component parts which might separately account for the relative influences of nature and nurture but which together could explain normal as well as pathological mental functioning. Though his model of the mind was more intricate than that of Descartes, Freud inherited from the Cartesian tradition a view of the mind as a private thought-theater, still operating in relative isolation from things in the world—despite his crucial discovery that the theater has a backstage (the unconscious). Much of the current criticism of Freud can be understood as a Hegelian response demanding that we look not just at how the mind's concepts contribute to our thoughts but how real things in the world (like perpetrators of incest and trauma) contribute to the shape of those concepts we apply in constructing our experience.

The Swiss zoologist, philosopher, and child psychologist **Jean Piaget** (1896–1980) was the first person to attempt a systematic study of how children *in fact* acquire the concepts we apply in constructing our experience of the world. By asking children ingenious and revealing questions about simple problems and by analyzing their age-appropriate but mistaken responses, he devised a picture of their way of conceptualizing the world. Like Freud, Piaget identified a number of discrete stages of mental development in children. But more than Freud, Piaget emphasized not only the way existing concepts are used to "assimilate" the world (as Kant said) but also the way these concepts "accommodate" themselves to the actual features of the world (as Hegel said). This two-fold process of assimilation and accommodation was explored by Piaget in a number of separate books on children's developing conceptions of time, space, and causality, among others.

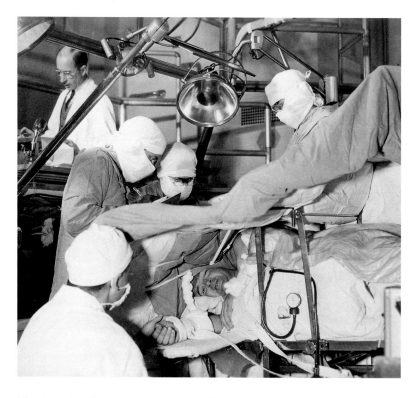

The American-born neurosurgeon **Wilder Penfield** (1891–1976) was a pioneer not only in neurosurgical technique but in his belief that neurosurgery could be used to make important advances in our scientific understanding of the brain. Penfield was the first director of the Montreal Neurological Institute, where he is here shown in the operating room (circa 1950) performing brain surgery on an awake, alert patient. Penfield was the first to ascertain the location and extent of brain lesions by opening the skull under local anesthetic and then using an electrode to stimulate the brain around the lesion. By asking the patient undergoing surgery what experience was being evoked by the electrode stimulation, Penfield could try to spare adjoining regions that proved to control crucial functions like language, vision, or memory. These direct observations suggested that the brain works in something close to Kant's model, with identifiable parts involved in initially analyzing sensory data and other parts involved in synthesizing this information into our experience of the world.

The neuroscientists **David Hubel** (left) and **Torsten Wiesel** (right) celebrate their 1981 Nobel prize with a lecture summarizing the studies of the visual system for which they received the honor. Hubel and Wiesel discovered how the brain's plasticity at the cellular level enables things in the world to write their qualities into the structure of the brain. Their landmark studies have become a paradigm for brain research, in which neuroscientists no longer study only how existing brain pathways assimilate the environment, but also how the shape of those neuronal pathways accommodates to that environment—the very dialectic philosophers have sought to guarantee the validity of knowledge. As Wiesel later opened his Nobel lecture in Stockholm, "We were interested in examining the role of visual experience in normal development, a question raised and discussed by philosophers since the time of Descartes."

The neuroscientist, biochemist, and immunologist **Gerald Edelman** (who received the Nobel Prize in 1972 for elucidating the chemical structure of antibodies) has developed an elaborate theory of neuronal group selection which describes how neuronal systems can indeed accommodate the shape of their circuits to the qualities of the external world. This kind of system produces not only neuronal maps of things in the world but maps of its own maps that can lead to *conscious* experience of those things. Edelman is shown here with NOMAD (Neurally Organized Multiply Adaptive Device), a robot capable of just such accommodation of its "brain" circuits to the features of its environment (such as the features of the block Edelman is holding). Experiments with NOMAD and its predecessor, Darwin III, have demonstrated that some values must be programmed into the system (in the case of humans, by evolution) in order for robots or humans to develop any concepts of the external environment at all.

The cosmology developed by the theoretical physicist **Stephen Hawking** is a prototype of conceptual accommodation, as his evolving ideas about space and time have found those subtle inconsistencies in our current knowledge that serve as metaphysical hooks to grapple onto a more complete understanding of the cosmos. His theory of exploding black holes combined relativity theory (the physics of the largest structures in the universe) with quantum mechanics (the physics of the smallest structures in the universe). In a modern version of Kant's idea that the laws of physics actually relate to *us*, Hawking has also elaborated on what cosmologists call the anthropic principle, one version of which points out that the laws of physics and the properties of the cosmos are constrained by our being here, since only certain such laws and properties could give rise to life.

The English pediatrician and child psychoanalyst **Donald Winnicott** (1896–1971) started his career by studying biology at the University of Cambridge because of an early interest in the work of Charles Darwin. Winnicott spent his long professional life working with small children and their mothers to explore questions about early concept development, but without ever losing sight of the importance of the mother-child relationship to this development. Emphasizing that it is not appropriate to talk of "just an infant" but only of an infant-mother dyad, Winnicott identified the *social* nature of even our most basic ideas about how the world works—a point of view that found its voice in philosophy with Wittgenstein's argument that a "private language" could not be possible, since even the most basic concepts expressed by language are irreducibly social in nature.

The physician-scientist **Jonas Salk** is best known for his development of the vaccine that saved untold numbers of children from crippling poliomyelitis. Salk's model of his polio vaccine is the addition of something artificial (a chemical potentiator) to something natural (the polio virus) in order to *improve* upon nature (prevent disease). This parallels Hume's model of morality precisely, wherein we mix our artificial values (like justice) with the world in order to make the world a better place. In his recent work, Salk has raised the question whether we are now moving beyond the simple mixing of our thoughts with nature (as in Darwin's artificial selection) toward a new evolutionary era in which we take a more active role in applying our thoughts to shape the course of evolution itself—with all the responsibility this would entail.

In pioneering studies of the heritability of human intelligence, the English explorer and anthropologist **Sir Francis Galton** (1822–1911) coined the term "nature versus nurture" for the relative contributions of heredity and environment. He conducted some of the very first twin studies and, with his student Karl Pearson, developed the mathematical tools of regression analysis that we still use to tease apart correlations. A cousin of Darwin, Galton was among the first to recognize the implications for humankind of the theory of evolution, and he coined the term "eugenics" for the movement he led to increase the proportion of persons with what he considered above-average genetic endowments through the selective mating of marriage partners. The history of the eugenics movement into the twentieth century highlights the many deep moral issues that permeate the nature-nurture debate. It is seen today in discussion of the ethical concerns raised by the Human Genome Project, ranging from the uses and misuses of genetic testing to the engineering of human genes in ways Galton could have only imagined.

11

The Plastic Brain

The Kantian philosophical tradition of the last two hundred years has so ingrained in us the idea that the structure of our thoughts contributes to our experience of the world that it seems almost intuitive that the structure of our brain contributes to our experience of the world. This is most obvious in the perceptual sensory realm where, for example, we humans only have direct visual experience of the spectrum of light between red and violet because of the structure of our eyes, retinas, optic nerves, and visual cortex. Researchers have shown that bees, birds, and other creatures have very different visual perceptions of the world, since their visual apparatus is constructed to register light waves outside our range. In this obvious sense, the structure of our nervous system shapes our experience of the world.

But since the brain is the organ we use to *understand* as well as to perceive the world, it has made some sense to look for brain structures that shape the more conceptual aspects of our experience as well. In the preceding chapter, we saw how neurobiologists have begun to discover brain areas that are responsible for the experience of objects as objects and for the spatiality of experience. When these areas become impaired (say, by a stroke or a tumor), a person's conceptual experience of the world is profoundly altered. Still within the Kantian tradition, it is therefore tempting to come to the conclusion that the reason we experience a spatial world filled with objects is because we have brain structures that construct any experience we can have by applying the inherent concepts of spatiality and object-ness to our sensory experience. If the story neurobiology can tell us ended here, it would in fact support Kant's unsettling conclusion that our experience records nothing more than features of the world-as-we-know-it, in distinction from the world-as-it-really-is, about whose spatiality and object-ness we can say nothing.

In this chapter, we shall explore the Hegelian rejoinder to this Kantian story in the idiom of the new neurobiology. I call this the "new" neurobiology because most of the exciting research of the last twenty-five years in the neurosciences has focused not on the me-chanical ways in which our adult brain structures set limiting condi-tions on cognitive processing, but rather on the subtle ways in which the experience of the environment in early life actually gives shape to those neural structures that will be used throughout adult life to construct our experience.

As early as the 1930s scientists had begun to suspect that the visual environment itself helps to shape the neural connections in the vis-ual cortex that in turn are used by us to perceive that visual environ-ment. This line of thought can perhaps be dated back to 1932, when Marius von Senden in Germany reviewed the world's literature on congenital cataracts. Children born with these opacities in the lens are deprived of the experience of patterns and shapes in the environ-ment but are still able to perceive light and dark. Von Senden discov-ered several children whose congenital cataracts were removed later in life. Presumably the retinas, optic nerves, and visual cortex of these children were perfectly intact. However, when their cataracts were removed between ages ten and twenty, the patients could recog-nize colors but had difficulty recognizing shapes and patterns, even after they were fitted with glasses that properly focused these envi-ronmental images on their normal retinas. The suspicion arose that the visual cortex of these individuals was not normal, presumably as a result of their childhood deprivation of the very shapes and patterns they were later unable to recognize. This was the first empir-ical hint that normal postnatal sensory experience of the world contributes to the development of the perceptual apparatus we use to have that very experience.

The hints from this natural experiment were supported by subsequent laboratory studies in which cats, monkeys, and other animals were raised in environments of sensory deprivation. For example, Austin Riesen in the 1950s raised newborn monkeys in complete darkness for the first three to six months of life. These monkeys could not discriminate even simple shapes when later introduced into a normal visual world.

During the 1960s and 1970s, more selective sensory deprivation of newborn animals also began to provide subtle behavioral clues concerning the contributions of postnatal visual experiences to the later capacity for normal perception of the environment. Investigators devised ingenious ways to rear newborn kittens in environments containing stripes of only a given orientation (vertical, horizontal, or oblique) or stripes moving in only a single direction. Kittens raised in horizontal environments for only three to four months and then put into normal environments were blind to other orientations but could see horizontal lines within the normal visual world. If, using goggles, one eye was allowed to view only horizontal and the other only vertical stripes from birth, the animal would later respond to one or the other (but not both) if one or the other eye was closed!

All of these studies confirmed the idea that early experience of the environment contributes to perceptual competence in later life. They also confirmed the existence of *critical periods* in the development of the brain—periods of enhanced "plasticity," as well as vulnerability. Visual deprivation in *adult* animals for even long periods of time had no effect on later perceptual competence. Presumably, the developing perceptual apparatus is sensitive to the effects of the environment only during critical periods of development, which differ in duration for different species and for different stimuli within the same species. Critical periods for specific kinds of learning are ubiquitous in animals. Sparrows need to hear and reproduce their parents' song by a young age or they will never get it right, and ducklings only learn to follow large moving objects in the first few days or weeks of life. While visual plasticity for stimuli like those described here (orientation, direction of movement) is thought to have a critical period of only three to four months in kittens, this critical period is known to last up to two years in macaque monkeys, and the human visual system may have a critical period as long as five to ten years, although most of the effect occurs very early in life.

Other investigators have found critical periods in the development of social competence as well as perceptual competence. The experimental and comparative psychologist Harry F. Harlow found that six to twelve months of social isolation in newborn monkeys led to severely withdrawn, asocial behavior, while a comparable period of isolation later in life was innocuous. Similarly, the studies of institutionalized human infants discussed in Chapter 7 revealed not only the cognitive impairments described there but also profound social disturbances related to severe depression only when their isolation occurred during the early years of life. As with the effects of later perceptual deprivation, social isolation later in life—whether because of imprisonment, hospitalization, sensory loss, or other reversible causes—has much less impact on subsequent social capacity (even if these prolonged, unpleasant experiences can lead to behavioral responses below one's capacity).

The intimate connection between plasticity and vulnerability during critical periods of development has many implications. We might, for example, be able to learn about the length of critical periods by studying the time periods during which various systems appear to be vulnerable. It is known, for example, that being born "cross-eyed," which causes the eyes to be unable to coordinate their movements and hence be unable to send the brain information about the same thing simultaneously, can affect the developing visual system in children until the age of four or five years. Requiring a single image of the world, the brain will soon phase out information coming from the slower eye—acting almost as if it had a patch over it. Treatments such as eye-patching the "good" eye for a portion of every day to enable the "bad" eye to set up proper cortical connections can have profound therapeutic effects if carried out during these critical years. If such patching is done effectively, then by the time these children are old enough to have a surgical procedure to repair their eye muscles, their plastic brains have wired themselves up to both eyes correctly. Thus, while the enhanced plasticity for which evolution selected us has its price, it also has important clinical implications.

Another excellent example of a critical period of development is human language acquisition. Like those young sparrows needing to hear and reproduce their parents' song to get it right as adults, the profound effects of childhood deafness on later language acquisition is well known—as is the lack of impact on linguistic competence if deafness occurs later in life when plastic language centers have already been formed. This goes not only for language as a whole but for specific elements of language. Kikuyu and Spanish infants easily discriminate the English ba's and pa's that Kikuyu and Spanish do not use and which the parents of these infants cannot tell apart. Similarly, Pinker describes how English-learning infants under age six months easily discriminate sounds used in Czech, Hindi, and other languages that English-speaking adults cannot hear, even with five hundred trials of training or a year of university coursework. The babblings of babies start out the same the world over (even if their parents are mute). But by about ten months of age, the human speech input centers have "tuned" themselves to the sounds of the language in their environment, so they can instead begin tuning into words and grammar—with the result that a one-year-old English-learner has already stopped distinguishing Czech or Hindi sounds that could be perceived just a few months earlier.

Although all of these behavioral observations support the idea that during early critical periods of development the nervous system is plastic, the belief that the actual physical connections of brain cells were being molded by the environment was still speculative. This began to change with the introduction of new techniques in cellular neurobiology that enabled researchers to study the environmental impact of early experience not just on behavior but on cellular development as well. These experiments are best exemplified by the original Nobel Prize-winning work of David Hubel and Torsten Wiesel.

Using microelectrodes to study the circumstances under which specific individual cells fire in the visual cortex, Hubel and Wiesel were able to study both the anatomy and physiology of the visual system in animals raised in states of selective sensory deprivation. Their findings were striking. The neuroanatomical structures normally devoted to input from perceptual features absent in the environment actually shrink at the expense of those devoted to input from features of the environment which are present.

For a specific example, let us return to the anatomy lesson of the last chapter. Recall that incoming information from each eye converges for the first time on incoming cells in the primary visual cortex. There, identifiable columns of cells each analyze some specific perceptual feature of the environment, such as the orientation of lines of contrast. In the 1970s Hubel and Wiesel carried out experiments on the plasticity of these orientation columns by raising animals in environments which selectively deprived them of certain orientations. In one such set of experiments, as mentioned previously, kittens were raised for the first months of life wearing cleverly devised goggles that only allowed them to experience horizontal lines. With no vertical or oblique lines in their environment tuning the orientation columns of their visual cortex, these kittens went on to be blind to all but horizontal lines when the goggles were removed after the critical period (which lasts three to four months in this case).

But more than demonstrating this behavioral observation, Hubel and Wiesel built upon their earlier elegant work, which had mapped out how all the columns of cells in a kitten's primary visual cortex divide up the environment's orientations among themselves. What they were able to show is that experience with horizontal lines in the environment during the critical period allowed horizontal orientation columns to develop in the primary visual cortex. Columns that would have come to represent vertical or oblique lines shrank in size or were switched over to represent the only orientation the world appeared to offer. This also works in the opposite direction for kittens raised in a strictly vertical or strictly oblique environment. With no horizontal edges contributing to the developing visual system, the adult cortex becomes wired up so the information about horizontal orientations cannot be experienced, these columns having shrunk or been replaced by some more "useful" orientation.

To summarize this set of experiments, Hubel and Wiesel found that, in essence, horizontal edges in the environment quite literally shape the development of horizontal-orientation columns. The experience of horizontal aspects of the external world in adult life thus depends upon the contributions of horizontal orientations in the environment to a plastic brain whose cells would have tuned themselves to other orientations if the external world contained no horizontal edges.

This pioneering work has since been extended not only to other aspects of visual cortex development (for direction of movement, color, binocular disparity, and so forth) but also to many other cortical areas (hearing, touch, and so on). It now appears that similar mechanisms as those found in orientation columns are at work in the shaping by the environment of the developing brain's ability to experience many properties of that environment. There is little reason at this point to doubt that at least some neural plasticity directs the organization of all the sensory input columns of cells in the cortex (a.k.a. the faculty of sensibility).

What was so striking about these experiments was not just the existence of the brain's plasticity, but the *degree* of its plasticity. For example, in studying the plasticity of orientation columns of kittens, Blakemore and Mitchell found that just *one day* of exposure to vertical stripes (on day 28 of the kitten's life in their experiment) was sufficient to bias the response of some neurons toward the vertical when the kitten was tested again after a further six weeks in the dark. When nature harnesses nurture to help orchestrate brain development, what gets harnessed is not taken lightly!

Having focused in the last chapter on the ways established brain structures *assimilate* the visual environment (to use Piaget's term for the slicing and dicing that takes place as we incorporate stimuli from the environment), we can now see the corresponding mechanism that nature has provided for the brain to *accommodate* itself to that environment, by changing its own structure in response to stimuli. Although Wiesel did not mention Piaget or accommodation as such in his Nobel lecture, the lecture began by placing this scientific work in the context that they were "interested in examining the role of

visual experience in normal development, a question raised and discussed by philosophers since the time of Descartes," and it concluded
with a language familiar to us from this cross-disciplinary perspective:

> Deprivation experiments demonstrate that neural connections can be
> modulated by environmental influences during a critical period of
> postnatal development. We have studied this process in detail for one
> set of functional properties of the nervous system, but it may well be
> that other aspects of brain function such as language, complex percep-
> tual tasks, learning, memory, and personality have different programs
> of development. Such sensitivity of the nervous system to the effects
> of experience may represent the fundamental mechanism by which
> the organism adapts to its environment during the period of growth
> and development.

In the decade that has passed since Wiesel's lecture put this scientific work in the context of our nature-nurture story, a huge amount
of effort has gone into studying not just cellular changes but the
actual *molecular* mechanisms that make possible these plastic
changes in the brain. Having already seen how important is the influence of the local internal environment on determination, differentiation, and embryonic maturation (as when the positioning of
future muscle cells near immature would-be skin cells causes these
immature cells to become nerve cells), we should not be surprised
to discover that the local cellular environment continues to play an
important role in the postnatal development of the brain. Indeed,
within each synapse (each cell-cell connection) and in identifiable
locations on each synapse, researchers have begun to uncover specific proteins that appear to be implicated in the brain's ability to
undergo plastic changes. Two that are worth mentioning are the so-
called NMDA receptor (a protein that sticks out of some nerve cells
to help regulate information from neighboring cells) and the
growth-associated protein GAP-43 (a protein implicated in neuronal
growth, maturation, and plasticity).

GAP-43 is a very interesting molecule in that it is produced by nerve cells in large quantities during the embryogenesis of the visual system (and, indeed, all other brain pathways), and it continues to be synthesized after birth for a period of time which appears to correlate with the critical period for visual development. Nerve cells then largely stop making it, and it never appears again in the mammalian visual system. The gene for GAP-43 does, however, get turned back on in those species where damaged nerve cells are capable of regenerating. For example, a goldfish optic nerve produces GAP-43 during its development and critical period and then shuts it off; but if the goldfish optic nerve is severed, GAP-43 is synthesized again throughout the time of its regeneration and reconnection of synapses and then turns off once again (lucky goldfish). Such observations have suggested that one requirement for neural shaping by the environment, whether during prenatal development, critical period plastic modeling, or regeneration, is the presence of a set of growth-associated proteins, of which GAP-43 appears to be the most important.

One reason it is worth focusing on the GAP story is that these growth-associated proteins drop out after the critical period in the perceptual areas of the human brain (the sensory cortical areas connected with vision, hearing, and so forth), *but they continue to be present throughout adult life in most of the association cortex and deeper limbic structures.* This important distinction between those brain areas which stop making GAP-43 after a critical period and those areas that continue making it throughout adult life is worth emphasizing. For one thing, it provides the further neurochemical evidence I promised in the last chapter for our map of brain regions. In addition to the anatomical and functional evidence already provided, one reason to believe that the human brain can be understood in terms of interesting component parts (or faculties) is that, in this case, the sensory input systems (sensibility) can only manufacture GAP-43 during a set critical period after birth, while in the association cortex and deeper limbic structures this crucial protein continues to play a role throughout adult life. With our anatomical background, the visual system can again illustrate this point. In studies of GAP-43 in the adult human brain, almost none of the protein can be found in the adult primary or secondary visual cortex, while a great deal is found beginning at the boundary with the visual association area, and its level increases quantitatively as we move farther into the deeper association cortex (where, for example, the objectness and spatiality of experience appear to be constructed).

This molecular, neurochemical distinction between cortical regions carries with it a profound implication. Since GAP-43 has been associated with the capacity of neural systems to reshape their connections in response to environmental input, the continued expression of this protein in our association cortex and limbic system raises the possibility that these brain regions are not limited to a critical period with respect to when in life the environment can contribute to their structure. Indeed, this is a likely possibility, based on animal studies of these regions of the brain. In one of the most highly developed experimental models of continued plasticity in these regions of the adult rat brain (a model known as "long-term potentiation"), investigators have found that a dramatic increase in the activity of GAP-43 occurs specifically when these *adult* synapses in the limbic system undergo plastic changes—changes no longer possible in sensory input areas where the genes responsible for GAP-43 have (for still unknown reasons) turned off permanently.

With this molecular evidence adding to the case for a division of brain regions into something like Kant's faculty of sensibility and faculty of understanding—in addition to our output faculties, emotional faculties, and so on—it becomes clear that we are *not* speaking metaphorically when we divide-to-conquer in our analysis of brain function. In fact, the empirical evidence used to justify our division of the brain into distinct parts turns out to have the greatest possible relevance to the nature-nurture debate. Here are genes that we inherit for the production of a protein that helps determine when and in which parts of the brain plasticity can allow the environment to play a role in wiring the neural connections we use to experience that environment.

Specifically how these growth-associated proteins and other molecules (like NMDA receptors) enable the cells under Hubel and Wiesel's microscope to accommodate themselves to the environment during the critical period is not yet known. What does appear to be the case, however, is that while those sensory areas of the brain responsible for perceptual competence (sensibility) can accommodate to the environment only during a critical period of development, the association and deeper limbic areas of the brain (understanding) may continue throughout adult life to have this plastic potential.

All of the philosophical and psychological analysis in the first part of this book pointed to this capacity of the human mind to remain fluid and dynamic throughout life in areas where ideas and concepts are formed. This was why Hegel spoke not merely of the "discovery" of knowledge but of the "becoming" of knowledge, as adult experience of subtle inconsistencies and inadequacies can lead to the development of concepts more consistent with broader contexts of experience. If you might pardon the hyperbole, we may well wonder whether GAP-43 is one of the "humors" Hegel might have sought had he explored the scientist's, rather than the philosopher's, view of the investigation of the "helping hand of things." Our Hegelian analysis of knowledge suggested that Kant's grid of understanding must remain fluid during adult life if even our basic concepts are to accommodate themselves to the actual world with which we interact. If that analysis pointed our philosophy toward the goal of discovering how "having a concept of a thing can be dependent upon things," it helps to know that our concepts about the world may be sculpted by the shape of the world as our conceptual understanding matures with ever-broader adult constructions of that world.

This neurodevelopmental perspective not only supports the philosophical argument that we have been building, it is also consistent with our discussion of stages of development in Chapter 6, but now in a more sophisticated way. We have discovered how a variety of "constancies" found in the external world determine the knower's ability to experience those constancies. In our twin experiment, it was the presence of objects in the world that provided the structure needed for minds to construct experience according to the concept of "object." Now we have seen how the presence of horizontal lines in the world provides the structure needed for us to construct experience according to horizontal features of the world. It is striking that a critical period exists only for the latter of these two constancies: for the presence of horizontal lines in the environment but not for the presence of objects. During critical periods in the development of sensory areas of the brain, nature presumably has taken for granted that all of the relevant perceptual building blocks will be present (including shape orientations, colors, depths, and language sounds), and the sensory system can "hook itself up" so that it can operate with great speed and efficiency thereafter. A faster, more efficient sensory system has a clear evolutionary value, since the senses must keep in touch with often threatening external events that would make it too dangerous for this system to slow down to draw on all our knowledge (which is why the Müller-Lyer arrows still *look* unequal even when we *know* they are equal). Higher-level conceptual features of the world, in contrast, may continue to refine themselves well into adult life, as would be required if knowledge is to become an ever-improving achievement along the lines discussed above. While we do not know if this plasticity of our basic concepts slows down—along with most other bodily functions—toward the end of life, a good friend and brilliant psychoanalytic scholar, Alfred Margulies, has suggested to me that we might think of old age as "hardening of the categories"!

What is most important of all here is that both the perceptual and conceptual features of experience (say, both colors and spatiality) have become understood developmentally as features of the *environment*—of the external world of things—which shape our "mental structures" so that we can construct our internal experience according to those external features. This is a long way from Kant's original view of these features as fixed properties of a mind which constructs the external world according to these properties that are strictly part of its own nature! All of our faculties—not only sensibility and understanding but also our output faculties such as movement and speech, and indeed reason itself—may now be understood as active faculties which interact not only with one another but with a world capable of writing its signature into our thoughts about it. Far from that nature-side story about evolution crafting genes to shape fixed brain structures for fixed conceptual features of the world, the nature-nurture mix described by the new neurobiology provides for an ever-improving conception of the world.

Such an ever-improving conception of the world was precisely Hegel's challenge when he spoke of using reason to search our basic concepts for inadequacies and inconsistencies that might point to larger truths. We have already seen some examples of this process. A simple example from the perceptual side is how we come to have knowledge of those wavelengths of light beyond our human visual spectrum. We interact with a world where the behavior of birds and bees makes no sense until science demonstrates how these creatures zero in on patterns in the flower not visible to humans; we soon devise machines to help us appreciate these parts of the spectrum more directly. A profound example from the conceptual side is Hawking's view of time. By searching our cosmos for subtle clues, we can come to have knowledge about periods of time when the universe does not even manifest living beings: what Russell described as Hegel's "metaphysical hooks" are now present in the universe, and we can use these hooks to grapple our knowledge to these other times, even if we will never be there to experience them directly.

The analogy between this simple example of our visual perception of the world and this profound example of our temporal conception of the world may open a new window on our entire nature-nurture discussion about the potential limits of human knowledge. Genes will inevitably be found that determine the structure of our retina, with its three types of cones that limit our direct experience to what we call the "visual" spectrum. But other genes have ensured enough conceptual plasticity in our brains that we can appreciate ultraviolet and infrared waves quite easily, and this appreciation has changed (broadened, *improved*) our understanding of the nature of light. On this analogy, there may well be genes that determine a deep universal grammar that similarly constrains our direct experience of time, but other genes clearly allow us to reach beyond this experience and appreciate other temporal realities—even other realities (like a contracting universe) that could never shape anyone's genes since they are incompatible with life. Perhaps these grammar genes only constrain our experience of time in the limited way those other genes constrain our experience of lightwaves. Maybe Chomsky and Pinker are right and Hopis do not experience time differently, because their grammar is genetically destined to have tenses like every other language. One thing I know for sure, though, is that *I* have begun to experience time differently since I started learning about the current scientific thinking regarding time's cosmological nature, so I know that in certain conceptual areas of my brain some plasticity exists with regard to the experience of time.

This example of how reason can search a basic concept like time to find broader contexts and improve our knowledge of the world still carries with it the notion that our conception of time is "better" if it now corresponds more precisely to the way time actually flows "out there" in the cosmos. Indeed, some thinkers have taken all the elegant results of the new neurobiology to support an elaborate theory of knowledge which relies on a one-to-one correspondence between our thoughts about the world and the world itself as the only way to support the veracity of our beliefs about that world. They suggest that perhaps all of the machinations of our faculties and their plasticity are there to ensure that this correspondence will give us an "accurate" picture of the world. As we shall see in Chapter 13, nothing could be further from the truth.

Before we reconnect our view of the new neurobiology with our earlier philosophical analysis, however, let us first turn briefly to values and feelings, and reconsider what was said about "feelings and things" in the light of our neurobiologist's picture of the mind.

12

Even Robots Need Values

With the burgeoning of neuroscience described above, several theories have begun to emerge about how the human brain comes to have conscious knowledge of the world. The most impressive of these is the elegant theory developed by the Nobel laureate Gerald Edelman, which he has called the theory of neuronal group selection.

It is one thing to understand how the visual cortex can construct a visual map of the environment and quite another to understand how the association cortex can construct a conceptual map of the world. In the former case, we have seen how early postnatal visual experience during the critical period shapes columns of cells in the visual cortex to organize around specific features of the environment, so that these perceptual building blocks can be used throughout adult life to construct an accurate picture of the world (more about "accurate" in the next chapter). If indeed the neurons in the association cortex retain some plasticity into adult life, as we have reason to believe they do, how do they pick out the conceptual building blocks from the environment that will likewise provide such an accurate picture of the world?

According to Edelman's theory of neuronal group selection, the processes involved are "selective" in something like the way the processes of evolutionary change are "selective." That is, groups of neurons are engaged in something like a biological *competition* in the process of their maturation and their forming connections to other groups of neurons. In this case, populations of neurons are "selected" through the strengthening of cells and connections that function repeatedly, and through the weakening of cells and connections that do not. In the visual cortex, a column of cells comes to represent horizontal lines because repeated firing in response to horizontal lines in the environment has stabilized those cells and their connections.

We have already seen how our relatively small number of genes could not be expected to provide a wiring diagram for the extraordinarily large numbers of cells and connections in our brain, but our genome presumably does set up the constraints that make this competition work by producing all the biochemical machinery (including molecules like GAP-43) to harness the (internal and external) environments to make it work. It has been almost fifty years since the pioneering neurophysiologist Donald Hebb demonstrated how the activity of one neuron on the next strengthens the connection between the two, and this has now emerged as a guiding principle of the competition and selection among neuronal connections in the embryonic and postnatal plastic brain: "Neurons that fire together, wire together."

According to Edelman's theory, what emerges from this competition among groups of neurons in, say, the visual system, can be thought of as multiple "maps" that are interconnected to one another through massively parallel and reciprocal connections. Although we mentioned a few of the visual maps above (for orientation, direction of movement, color, binocular disparity), over thirty interconnecting maps have been identified in the monkey visual cortex.

The competition among groups of neurons becomes much more complicated when we add in these multiple interconnections between maps, since the process of some groups of neurons being stabilized through repeated functioning in one map will also stabilize the groups of neurons in connected maps which are also functioning as a result of these connections. Edelman uses the term "re-entrant" to describe these circuits through which the brain's own maps "talk" to one another and can alter and reinforce their connections and wiring in the process.

If Edelman's description of how such perceptual maps emerge is true (and there is already good evidence for it), one can immediately ask whether a sufficient number of these perceptual maps could together account for the development of *concepts*. Since we are talking about one system (the brain) with presumably one set of rules (complex as they may be) governing its biological function, perhaps we should look no further for the ways *conceptual* features of the world write their signatures into our brains?

The reason it has to be more complicated in the "faculty of understanding" is that concepts are enormously heterogeneous and general, as they involve mixtures of relations concerning both the real world and past memories and behaviors. In Edelman's words, "Unlike the brain areas mediating perceptions, those mediating concepts must be able to operate without immediate input."

It is therefore no small matter that most of the brain gets its input from, and gives its output to, *other parts of the brain* (and not the outside world). Most of the brain spends most of its time trying to figure out what is going on *in the brain!* The theory of neuronal group selection suggests that, while the perceptual areas of the brain use their re-entrant circuits primarily to construct maps of *external* stimuli, the conceptual areas of the brain use their re-entrant circuits primarily to construct maps of the brain's *own* activities.

In fact, with the benefit of global mappings developed in each of the sensory modalities, brain areas responsible for concept formation can categorize, discriminate, and recombine the brain activities occurring in different kinds of global mappings simultaneously, and also categorize, discriminate, and recombine all of these with *past* global mappings in each modality as well as with the presence or absence of past movements, relationships between perceptual categorizations, and so on. Conceptual brain structures must represent "a mapping of types of maps," as Edelman puts it.

Some of the neural structures capable of such activity are the frontal, temporal, and parietal association cortexes of the brain. We already observed that, under the lower magnification of our microscope, these association areas look much more tangled and interconnected than the more hardwired appearance of sensory input systems, with their discrete, identifiable pathways, and we can now see why. Their massively re-entrant circuits must be able to activate or reconstruct present or past (or both) activities of mappings of different types and be able to recombine and compare them. They must be able to store and retrieve memories of other brain activities and operate in many cases independently of current sensory input.

Edelman's neurobiological story of how concepts arise in the brain has many consequences for all we have discussed thus far. As a historical point, it certainly highlights the remarkable insight of some of the thinkers we have already looked at, who could only deduce how the brain arrives at its concepts through philosophical introspection or through very gross behavioral observations of infants and children. Recall, for example, Piaget's description of the accommodation of our concepts to the environment, as when the preoperational concept of object emerges through "the child's network of inter-coordinated sensorimotor schemas" as the infant begins to "appreciate that the rattle she sucks is the rattle she sees and the rattle she grasps." Here Piaget is describing how repeated (re-entrant) comparisons between multiple present and past perceptual maps can give rise to a new conceptual understanding of the environment. This picture of concept development likewise highlights Hegel's remarkable insight in describing the process by which concepts continue to accommodate to reality throughout adult life, as we "search our concepts for the inadequacies and inconsistencies that might reveal broader contexts of experience" (continually recomparing and recombining our maps of our maps in plastic brain areas).

But Edelman's own insights into how concepts arise in the brain do much more than give us a deeper appreciation of the power of philosophical or psychological analysis. Let us consider the implications of the theory of neuronal group selection for some of the issues raised thus far concerning knowledge (as well as some issues soon to be addressed concerning values). Certainly the first significant consequence of this view of concept development is how far it brings us from Descartes's totally and Kant's partially passive view of the brain.

Since it was this passivity that led to Descartes's radical—and Kant's relative—independence of thoughts from things, our current emphasis on the *activity* of the brain (in sensibility, understanding, and elsewhere) can perhaps help us finally reunite thoughts and things and, as Hegel suggested, make knowledge not merely an activity but an achievement.

Indeed, although we have here a divide-and-conquer approach to understanding how knowledge develops (with separate input analyzers, conceptual processes, output systems, and others we shall discuss shortly), for the first time we divide-to-conquer in a way that does *not* automatically leave nature in domination. Although "brain structure" has traditionally been taken as part of our "natural endowment," our new neurobiology paints a picture of highly interconnected subsystems (with feedback as well as feedforward connections), almost all of which are plastic for at least some critical period in their own development. So, although advances in science have generally fueled arguments for the nature side of the nature-nurture debate, the advent of this new neuroscience appears to be an exception.

This new neurobiology—with its demonstration of how our "natural endowment" has adopted the strategy of harnessing nurture through neural plasticity (combined with basic motor reflexes)—is not merely a *reflection* of the current pendulum swing back toward the nurture side; it is the engine driving that swing. This is clear to me not just as a student of the history of science and a reader of philosophy but as a parent and a reader of the popular "parents press" which, in the wake of all this research, is once again emphasizing the need to play Mozart to your developing baby in utero. Sales are soaring for curved horns through which mothers can "talk to their babies" during the middle trimester, and the linguistic research described above has sent some parents shopping for tapes of Czech, Hindi, and other such languages so their babies will not "miss out on the opportunity" to discriminate those speech sounds that become lost if the brain does not tune into them in the first ten months of life!

Perhaps I should clarify this point. I noted in Chapter 1 that we have recently been in a nature phase, with all my psychiatric training to reassure those concerned parents that their son's mental illness is the result of a "biochemical imbalance completely determined by genetics." Our new understanding of the brain and its development is sending us careening back to the nurture side, but it is a new kind of nurture argument that is much more synthetic of nurture *and* nature. We continue to discover the biochemical bases of the mental illnesses, but we presume that in most cases it is a "biochemical imbalance *not* completely determined by genetics." The psychiatrist now has a harder job—explaining how the biochemical imbalance likely arose from the influence of the environment on a plastic brain system that probably had some genetic propensity to be vulnerable to this environment. But we can be grateful for that difficulty if we are finally getting closer to the truth of the matter.

Edelman has also described several other implications of the theory of neuronal group selection that are relevant to our analysis thus far. As an evolutionary matter, for example, the theory implies that concept development—the ability to think—arose in the human ancestral line before language development (which Piaget showed is also true in each individual's development). Indeed, Edelman emphasizes how unlike elements of speech concepts are, since they do not require linkage to a speech community and (at least our basic concepts) are *not* arbitrary conventions—an important point also made by Pinker, to which we shall return. The priority of concepts over language is one of the reasons why I have not emphasized the linguistic approach to philosophical questions that has gained such popularity in recent years. Instead of moving from the likes of Kant and Hegel to Piaget to neuroscience researchers, this more popular approach would have moved from the likes of Wittgenstein to Chomsky to contemporary linguistics researchers. In so doing, most of the argument thus far would actually be the same. But the capacity to apply experience-shaping concepts in interacting with the world arises in each individual—and arose in our species—before the capacity for language.

This is not to detract from the important contributions of linguistic philosophy, since the ability to manipulate symbols (including words) unquestionably adds exponential power to the brain's ability to form maps of its maps. It is just that there is no neurobiological reason to assume that these interconnecting representational maps are limited by words as the only symbolic forms available. Indeed, there is even some evidence to believe that—far from language being what constrains human conceptualization—children may be able to learn language only after they have first learned how to make sense of the world, particularly of situations involving human interactions.

In George Lakoff's *Women, Fire, and Dangerous Things: What Categories Reveal about the Mind* (1987), for example, we get an interesting reversal of the constraints imagined by some linguistic philosophers. Lakoff describes how language is based on cognition, not cognition on language, once cognition is understood in terms of neurobiological functions. Within this same biological view, Edelman has even proposed a theory of how *self-consciousness* can arise through the process of continually comparing, categorizing, discriminating, and recombining current and remembered mappings in the brain in a way that certainly uses words to increase the power of this process, but with words being only one type of symbolic concept among many that the brain employs in this process.

Let me conclude this chapter with two important points about the theory of neuronal group selection and its implications. The first is about the nature of these concepts that are generated by our brains. The second is about the role played by feelings and values in this whole process.

If all of the neuronal events described above are selectional ones, arising from the relative stabilization of functioning synapses as a statistical matter (synapses can be relatively stabilized with some use, drop out with no use, or strongly connected with constant use), then we can see why the resulting wiring diagram of the billions of neurons in each brain must be different between all people, even between identical twins. But "selective" processes impose commonalities as well as variations. Just as the fine details of physical anatomy are different between people while any male and female pairing can engage in a common procreative process (one definition of species),

Edelman emphasizes that the selection of patterns of neurons results not only in individual diversity in the fine details of neuroanatomy but also in a "common pattern" in the species.

We can see a vivid example of this in the study of birdsong, where it is only the repeated act of hearing the species-specific song that enables each bird to select a (unique) neuronal wiring that enables it to sing the song, while this same love song is a species-defining part of its mating behavior. Just because all the sparrows in the species sing the same song, we need not postulate song-determining genes, since a more subtle genetic mechanism is enabling any birds that will survive and mate to accommodate this same song from the *environment*.

As we saw above, ultimately both twins (and all people reared in an environment of fairly permanent objects, numbers of things, and so on) will self-generate common concepts that arise out of unique neural connections in each. While this may not be true for all concepts (my concept of good music may not be shared by many), it can at least be true for the most pervasive constancies manifested by the world. Following this line of reasoning, we may end up with a host of "overlapping spheres" of groups of individuals participating in the same concepts—some spheres including very small groups but some including the whole species. We shall return to this point in Chapter 15.

Finally, let us consider the most important feature of the brain left out thus far—left out in part because it was "covered up" by the cortex during the late stages of brain evolution. Consider two important quotations from Edelman to which we shall continually return throughout the remainder of this book. First:

> While perceptual categorization and memory are necessary for learning, they are not sufficient. What is needed in addition is a connection to value systems mediated by parts of the brain that are different from those that carry out categorization. The sufficient condition for adaptation is provided by the linkage of global mappings to the activity of the so-called hedonic centers and the limbic system of the brain in a way that satisfies homeostatic, appetitive, and consummatory needs

reflecting evolutionarily established values. These value-laden brain structures, such as the hypothalamus, various nuclei in the midbrain, and so on, evolved in response to ethological demands, and some of their circuits are species-specific.

And second:

No selectionally based system works value-free. Values are necessary constraints on the adaptive workings of a species. In our species, the commonalities of physiological function, hunger, and sex imply a set of mutually shared properties. The brain is structured so as to play a key role in regulating the evolutionarily derived value systems that underlie these properties. Undoubtedly, these value systems also underlie the higher-order constructions that make up individual aims and purposes. We categorize on value.

What exactly is meant by this idea that "we categorize on value"? The implication is that our focus only on cortical brain processing has been quite inadequate. When researchers study the organization of concepts in the association cortex of the frontal lobes that appears to be so crucial to planning and complex functioning, they study not only the multiple re-entrant connections sent to and from other cortical sensory and conceptual maps but also the connections sent to and from those deeper brain structures that are involved with feelings and with determinations of whether physiological needs are being satisfied. When these more primitive brain structures sense a state of dehydration, their connections to the association cortex will, as it were, make us see a glass of water differently.

On what basis does Edelman claim that these primitive feelings are necessary for any conceptual categorization we carry out? Although we have some indirect neurobiological evidence—and we have Aristotle's insight that "reason by itself moves nothing; it is only when it is in pursuit of an end . . . that it moves anything"—we also have exciting new experimental systems that have been created to test various assumptions of the theory of neuronal group selection. The most famous example is Edelman's own simulated automaton named Darwin III.

Darwin III is a complex recognition automaton simulated by a supercomputer that can stabilize its "neuronal connections" through use, as its movable eye and four-jointed arm interact with the environment. Darwin III has a sense of touch at the last joint and a joint sense that tells it where each of the four joints is positioned at any time. Its program simulates some likely evolutionary pressures. For example, although no specific behaviors are programmed in, there is a program which values light falling on the central part of the eye. That is, Darwin III's program enhances the probability that synapses active when light stimulates the eye will be strengthened in preference to competing synapses, and even more so for light at the center of vision in preference to the periphery.

Programmed only with "evolutionarily set desires" or values such as light-is-better-than-no-light and touch-is-better-than-no-touch, Darwin III is put into an environment of objects—some of which are bumpy, some of which are smooth, some striped, some not striped. It starts by randomly flailing its arm and disjunctively sampling signals in each of its sensory modalities (vision, touch, joint sense). Darwin III's own interactions with the environment reinforce and inhibit further connections (selection), so that it activates its own re-entrant maps and soon discriminates with its flailing response between those objects that are striped and bumpy and those that are striped *or* bumpy but not both! What is remarkable about this categorization of the things in its world is that it occurs solely based on experience and not on the basis of programming. In essence, Darwin III uses the computer equivalent of neuronal group selection to accommodate primitive preoperational schemas from out of its initial sensorimotor schemas!

But Darwin III generates such primitive shape categories only if *some* basis for value is programmed in (light-is-better-than-no-light, touch-is-better-than-no-touch). Without *some* value programmed in, no categorization of the environment emerges. What "evolution" (in the form of experimental scientists) needs to create, then, is a being that "likes" to see and touch. The *being* will generate categories of rough versus smooth, striped versus plain objects so long as it interacts in multiple modalities with a world of objects both rough and smooth, striped and not. In this experiment, Darwin III does not generate concepts of sharp versus blunt or soft versus hard because these are not built into the environment with which it interacts. Again, we might say that, although its evolution did not code directly for any of these concepts—and the detailed circuitry that arises in each experimental run of the system is absolutely unique— the shape categories arising in Darwin III are "contributed" by the shapes in its environment.

Edelman's inference from this model is that both experiential selection of neuronal groups and some value-based circuitry are necessary conditions for the emergence of even basic concepts about the world. In his words,

> Categorization is not the same as value, but rather occurs *on* value. It is an epigenetic developmental event, and no amount of value-based circuitry leads to its occurrence without experiential selection of neuronal groups. But it is also true that without prior value, somatic selectional systems will not converge into definite behaviors.

To return to the human brain, we might conclude that the linkage observed from the frontal association cortex to the limbic system and other deep structures that determine whether physiological needs are being satisfied is a linkage that connects the two necessary (and, with experiential selection of neuronal groups, jointly sufficient) conditions for concept development. The deeper, value-laden brain structures (hypothalamus, various midbrain nuclei, and so on) set the physiological constraints for our homeostatic, appetitive, and consummatory needs. As the original engines driving the processes by which we categorize the world, they are presumably the evolutionarily established "values" that must have important implications for the sophisticated value categorizations we make in our ethics, however far our more subtle moral concepts are from these basic biological value systems.

In re-reading the two long quotations from Edelman above, we can now see the deeper implications of the conclusion "We categorize on value"—implications that have received further support from more recent experiments with Darwin III's successor, an actual robot named NOMAD (Neurally Organized Multiply Adaptive Device). When we turn to ethics, we shall reexamine this connection between the biologically based values upon which evolution has operated (surviving-is-better-than-not-surviving) and the highest order moral judgments we make.

The deep brain systems that control our feelings and basic physiological needs project their connections to widely distributed parts of the cortex and release substances that can modulate changes in synaptic strength. The psychoanalyst Modell could not have been more correct in his observation that "the capacity to know and the capacity to love are not . . . entirely separate functions." Indeed, we now have deeper insight into how brain functioning might connect the development of the object concept with the infant's feelings of discomfort in being cold, wet, and hungry during those minutes of waiting for the mother to appear. We saw how Blatt and Wild described those sequences of frustration and gratification as "primary experiences" in the development of object and self concepts. We can now understand this process as the way our *value-faculty*—programmed to set the physiological constraints for our own homeostatic, appetitive, and consummatory needs—interacts with our faculties of sensibility and understanding to give rise to our object and self concepts as we interact with the world.

We shall return to the interaction of values and concepts shortly when we elaborate a view of ethics that grows out of all our previous philosophical, psychological, and neuroscientific explorations. But first let us return, as promised, to look at what actually supports the truth of our beliefs about the world, since this will necessarily support the truth of our moral beliefs as well as the truth of our knowledge.

13

But How Can You Be Sure of That?

As soon as we recognize the logical distinction between thoughts and things, it becomes tempting to say about thoughts that they are *true* when they "correspond accurately" to the things they are "thoughts of." I have a "true" perception of a straight stick when I see it as a straight stick, and a "false" perception when I see it as bending at a pond's surface. With something like this correspondence view of truth Descartes was able systemically to doubt whether *any* of his thoughts corresponded accurately to things in the world.

Such a radical form of skepticism about the truth of our thoughts, about the veracity of our experience of the world, can always follow from such a correspondence view of truth. In order to see how correspondence theories of truth always leave room for skeptical doubts about such matters as whether everything we know might be false, or whether dogs and cats really exist, we need only consider how we go about *justifying* our thoughts. Consider, in other words, how we respond when asked to validate any idea or practice in general. The answer is usually simple enough in that we turn almost instinctively to the accepted standards employed by some relevant institution. To support our ideas about the day's events, we might turn to newspapers. To support our scientific ideas, we might turn to observations and experiments. To support our claim about the current score in a baseball game, we look at the scoreboard.

Sometimes the problem is more complicated, as when a legislator is asked to justify his or her ideas about housing reform or a scientist is asked to justify his or her ideas about how a certain metal will perform as a protective casing for spent nuclear fuel. But such individuals will still refer directly or indirectly to the standards of their respective disciplines, using a web of facts and procedures established and refined over time by many hands. Through sometimes very subtle use of this background of data, concepts, investigative techniques, and procedures, the collective knowledge of the relevant discipline is used to justify individual cases. It may sometimes be difficult to see how a (or even which) particular institutional background is to be applied, but in the vast majority of cases, the problem lies in the application and not with the institution itself.

There are, however, a small minority of cases when questions are raised about the fundamental nature of the *institution* whose practices embody the standards by which individual cases are to be justified. When continually asked for further validation, when pressed with round after round of "But how can you be sure of *that?*" we eventually come up against the foundations or underpinnings of our institutions—those fundamental assumptions which serve as the final court of appeals for problems within the institution. When pressed further for *what underpins the foundations,* we are left with a serious problem.

There are two basic approaches to this foundational problem. The first we can call an *internal* solution since it insists that we continue to remain within the institution in question. We call the second an *external* solution since it maintains that we must look outside the institution if we can ever hope to validate its foundations. The latter of these two approaches is in some ways the most natural. Common sense tells us that it would somehow be circular to seek support for foundations inside an institution supported by those very foundations.

While external solutions have the unreflective support of common sense, they also have problems. What standard is to be applied if standards are usually dictated by the institution in question? The answer to this invariably comes in the form of a search for some sort of fixed, indubitable, incorrigible, universal, and absolute principles that could serve as the ultimate final court of appeals for the basic questions. It is easy to see why these principles would need to have all of these wonderful qualities if they were to carry out their noble job.

Unfortunately, after centuries of searching, these absolute principles are still not forthcoming. In the case of the philosophical effort to secure the foundations of "true" human knowledge, the results have been downright abysmal. From Plato's forms to Descartes's principles of light, philosophers through the ages have futilely struggled to dig down to a bedrock layer of knowledge that is so incorrigible it can secure the rest of the edifice. But outside the institution of knowledge, there is no suitable place to dig. It is no wonder then that Descartes's ultimate standard of knowledge had to be the generous guarantees of a beneficent God and Kant's standard had to be an eternally unknowable world of things-in-themselves. Both were looking *outside* the institution of knowledge to underpin the foundations.

What were above called "correspondence theories of truth," then, are simply theories which insist that the validity of our knowledge is to be judged by seeing how it compares to some external standard (here, external reality). We can see why such correspondence theories are bound to lead us to the possibility of radical skepticism—to the ever-possible doubt as to whether we have *any* "true" knowledge—since external solutions presuppose that our standard for truth lies outside the boundary of knowledge itself. Once one accepts the idea of a boundary between the "inside" and the "outside" of knowledge, then crossing that boundary while remaining always on the "inside" (within knowledge) becomes impossible.

Common sense, in choosing against the presumed circularity of internal foundational solutions, thus leads us to the skeptical position, which itself has to be rejected by the same standard: common sense. Indeed, common sense ultimately supports the rejection of external foundational solutions not only because common sense supports the rejection of the skepticism which follows from their correspondence theories of truth. If we can at best begin "at the beginning," it only makes sense that we look *within* knowledge to understand how the edifice is supported. The fruitlessness of all those years philosophers have spent searching for an ultimate court of appeals *outside* of knowledge is not actually the least bit surprising. If we are looking to hit bedrock, then the only place to dig is within the institution of knowledge.

In the context of our neurobiological version of this story, correspondence theories of truth also make no sense because it is not at all clear where in the brain there ought to be an "image" that "corresponds" to the external world. The human retina and visual system is not a mirror designed to reflect the world "accurately" (if perhaps upside down and backwards!). On the contrary, it is designed to disassemble crucial features of the visual environment for the purpose of reconstructing an understanding of that environment which may well be called "accurate" but not because of how directly it has "copied" the world.

When we look at the human brain and how it does *in actuality* go about justifying its knowledge claims, we are not surprised to find that our own tests for validity do not involve a comparison of our knowledge with something external to it. What we in fact do is something like Edelman's description of a process of multiple re-entrant comparisons and recombinations of current and past global mappings in multiple modalities across multiple brain systems. As my close friend and philosophical sparring partner Thomas Steindler has put it, "We use knowledge that we have to test what we take ourselves to know." Just as the historian Thomas Kuhn demonstrated that scientific revolutions can shake the foundations of science without ever leaving the institution of science (á la Copernicus or Einstein), there must be some way that our knowledge itself serves as the final court of appeals for adjudicating the foundations of knowledge.

But what is it that replaces the simplistic correspondence view of truth once we reject the possibility of external solutions? What sort of standard can exist *within* knowledge that avoids the problem of circularity? The answer that comes out of the Hegelian philosophical tradition is that our ultimate standard must be *coherence,* and so we speak of a coherence theory of truth.

It is important not to take the term "coherence" at face value.

Coherence does not merely mean that we measure the truth of a given idea by testing whether or not it fits neatly into the background of all our other knowledge. Coherence, in Hegel's term, is not a matter of conforming to any single standard, even a standard as imposing as a relationship of coherence with every bit of knowledge ever known by humankind. For even *this* standard may prove inadequate as our power of reason struggles to find an ever-broader all-encompassing context within which to reconcile apparent contradictions we may find in our experience. When Hegel insisted that knowledge *becomes,* he meant to imply that not only knowledge, but also its criterion for validity, is dynamic, ever growing, developing, maturing.

We have already seen any number of examples of how knowledge *becomes.* At the level of individual development, each child's conceptual grid accommodates itself to the structure of the world in a series of developmental stages. The "coincidence" that those clay snakes are longer but also thinner than the clay balls they were rolled from dissolves as the child's conceptualization of the world broadens to encompass a rudimentary understanding of mass conservation. Once this stage is reached, we hold children to this standard, and when they make inaccurate statements about it, we no longer laugh at their cute preoperational logic—we tell them they are wrong, since the standard we apply to their knowledge has broadened along with their conceptualization of the world.

At the adult level of conceptual accommodation, we have also seen how even the shape of a basic concept like time can accommodate itself to the complex nature of the cosmos, as subtle inconsistencies in more primitive ideas about temporality can give way to a more sophisticated understanding of how events in our universe flow. Interestingly, once even a few scientific authorities have constructed a more sophisticated view of time, the more primitive views of the rest of us are not laughed off as "cute" but are considered outdated and, ultimately, just plain wrong. The dynamic standard by which all knowledge is judged has grown, developed, matured.

This brings us to the societal level, where the impact of scientific revolutions becomes absorbed and the standard by which all knowledge is judged also *becomes*. While Descartes was writing his *Meditations* in the first half of the seventeenth century, Galileo Galilei (1564–1642) was writing his *Dialogue on the Two Chief World Systems*, demonstrating how subtle inconsistencies and unexplained "coincidences" in the geocentric world views of Aristotle and Ptolemy pointed us instead to the Copernican view, which put the sun, rather than the earth, at the center of the solar system. When this broader context of experience became the new standard by which beliefs about the planets were to be judged, this standard applied to all human knowledge about the solar system. If explorers get off a boat and find themselves faced with a less developed culture where the people believe the sun revolves around the earth, we consider that they are wrong in their understanding of the matter— even if we can agree with how the sunrise and sunset appear to us each day even after a more all-encompassing context has been accommodated by the mind.

By introducing the way some people's knowledge of the world can change the standard by which the truth of other people's knowledge is judged, we inject an element of power politics into a philosophical discussion about knowledge (why do we call that culture "less developed" anyway, when the people there might have a much more sophisticated understanding of the human spirit than we do, even if they still believe the sun goes around the earth?). We are also reminded how important the role of *other thinkers* was in the earliest development of our concepts about the world. At the end of Chapter 7, we saw how even our basic object concept had contributions not just from "things" but also from "other thinkers." We saw how Winnicott's reminder that there is "never just an infant" broadened our picture of events in the nursery, so that we could no longer speak of "just an individual subjective experience," but only of the collective, *intersubjective* experience within which concepts of the world develop.

The participation of other thinkers in the nursery in our most primitive concepts of the world and the participation of other scientists years or miles away in our most sophisticated concepts of the world both teach us about the ever-expanding breadth of our criterion for the truth of our knowledge. Coherence must be sought not only within our *own* mind—searching our *own* concepts for inadequacies and inconsistencies that can point to broader contexts of experience. Coherence must also be sought within an intersubjective dialogue continually carried on with other thinkers, generated by a shared struggle for these larger contexts of the truth. We thus shift dramatically the philosophical rules of the game by changing Descartes's question "But how can *I* be sure of that?" and introducing someone else who is asking "But how can *you* be sure of that?" In so doing, we are no longer locked into the Cartesian trap of using the individual's *subjective* experience as the yardstick by which to measure the veracity of beliefs, by which to determine the *objectivity* of knowledge.

It was perhaps misleading then to use the metaphor of striking bedrock. As knowledge advances, it becomes possible that what we thought secure might need to be replaced. The brilliant modern philosopher Ludwig Wittgenstein instructs us to think instead of a river and riverbed. While our ideas (the river) move forward, they are supported by the bed of the river; but it is a bed both of rock and of sand. Certain beliefs, such as the geocentric solar system or the Euclidean nature of physical space, may be part of the bed—part of the "presumed foundations"—but later come to be swept away by the progress of knowledge, just as the sand of a riverbed is swept away by the movement of the stream. Here we see a certain historicism to even the foundations of our knowledge. Thus, the coherence theory of truth not only enabled Wittgenstein to validate in 1950 in his essay "On Certainty" the truth of the statement that "no one has been to the moon"; it also enables us to validate ever since July 1969 the falsehood of that same statement.

Within our scientific idiom, the apparent circularity of a coherence theory of truth is revealing, since we may likewise ask the neuroscientist, "But how can *you* be sure of that?" As biologists study the world at the level of organisms and systems of organisms, they eventually home in on cellular mechanisms that are themselves a function of the biochemistry of the molecules of which these organisms are made. Biological "truths," it might be said, must ultimately be supported by chemistry.

The Wheel of Knowledge

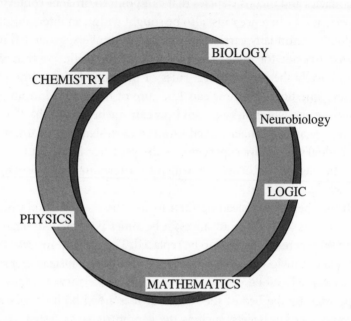

In a similar way, the chemist ultimately looks to physics for support as complex chemical interactions are eventually understood in terms of the forces acting upon the particles that make up all matter. The physicist, in turn, looks to mathematics to support the equations and laws that describe the forces of the universe. Indeed, even the mathematician looks further, to logic (a branch of philosophy!) to support the basic structure of mathematics.

We may therefore ask: What can do for logic what logic does for mathematics? Or what mathematics does for physics? Or physics for chemistry? Or chemistry for biology? What can support this most basic "science of reason"?

The circular answer that is part of our coherence theory of truth is that we ultimately depend on *biology* to support the laws of logic. If all of our above description is correct about the complex interrelationships of neurobiology and the external world, it is perhaps more specifically our *neuro*biology which supports our mind's use of logic. Those readers unfamiliar with basic mathematics may see no need to look for any further support of the basic laws of logic (two opposites cannot both be true; if A implies B and B implies C, then A implies C; and so on). But, as the early twentieth-century logician and philosopher C. I. Lewis put it:

> The ultimate criteria of the laws of logic are pragmatic. Those who suppose that there is, for example, *a* logic which everyone would agree to if he understood it and understood himself, are more optimistic than those versed in the history of logical discussion have a right to be ... Over and above all questions of consistency, there are issues of logic which cannot be determined—nay, cannot even be argued—except on pragmatic grounds of conformity to human bent and intellectual convenience.

Putting this into our Hegelian terms, we might say that we search through even competing and successive systems of logic for inconsistencies and inadequacies that reveal ever broader contexts of experience. We include in our logic that opposites cannot both be true because, in our interactions with the world, it *works* for us.

Thus, when we say that logic is itself ultimately supported by biology—and particularly neurobiology—we are using the term "support" in just the way chemistry supports biology, physics supports chemistry, and so on. Each level is necessary but not sufficient for the next: each "higher" level is grounded in and so must in some sense be compatible with its supporting level, but it is not reducible to that level, since new properties emerge at each. We are, in other words, saying that the only sense in which we can assert that it is a "necessary truth" that opposites cannot both be true is not a sense like Kant's (a priori) *logical* necessity but in some new sense of a *biological* necessity. While Kant was insistent on the existence of a priori logical truths whose necessity was completely separate from the contingent world of things, we have now come full circle from the a priori to the actual biological world. Within this circle, we discover the true depth and breadth of coherence which establishes the truth. Hegel told us that reason (including the laws of logic) are "relative to reality." We have now further specified that we are talking about *real* reality as it actually evolved in all its biological wonder. It is this biological world of things that has been implicated in our thoughts about it—even in the a priori concepts we use to construct any understanding we can ever have of it.

Let us consider some of the implications of this "wheel of knowledge."

14

Breaking Down Old Distinctions

The above considerations have blurred many of the distinctions historically thought to be crucial to philosophy and the natural sciences alike. We might begin with the distinction that is traditionally made between the a priori and the a posteriori. By definition, an a priori truth cannot be found to be true or false through experience since it is an essential feature of any experience we can have. Kant thus considered his categories of the faculty of understanding to be a priori features of experience, since in the necessary application of them in constructing our experience, we could not look further to any empirical evidence to determine their validity.

The distinction between the a priori and the a posteriori has been central to the whole problem of whether basic concepts can be learned: these basic concepts (such as number or object) would become "merely contingent" features of our experience if they were learned a posteriori (like the capitals of the fifty states), and philosophers have traditionally claimed a stronger a priori "necessary" status for these pervasive features of our experience of the world.

In our synthetic biological idiom, however, the traditional distinction between a priori and a posteriori has been blurred. Whether we have talked of Hegel's view of the becoming of knowledge, Piaget's view of the accommodation of concepts, or the neurobiologist's view of the plastic development of our association cortex, we have seen how those concepts we apply a priori in constructing our experience have been contributed in no small part by our interactions with the physical a posteriori world. In this model, a priori means "prior to experience" only in the sense that we each individually do apply these concepts in constructing our adult experience, but we can no longer presume that these a priori concepts are *independent* of experience. Our a priori concepts are therefore "merely contingent" features of experience to the extent that they are "merely contingent" features of the world in which we live.

This blurring of traditional boundaries is equally true for the distinction between the "form" and the "content" of experience upon which Kant based his theory of knowledge. Kant viewed the a priori concepts applied by the faculty of understanding as dictating only the form of experience, while the external world of things contributed only their "sensory contents." If that external world of things does indeed also write itself into our faculty of understanding, then even the form of our experience has within it the signature of the helping hand of the a posteriori world of things. Indeed, the sharp distinction Kant made between the perceptual building blocks (shape, color, motion) and the conceptual building blocks (object, space, time, causation) which we assemble together to construct our experience is a distinction that may itself be subject to a posteriori empirical investigation once we take seriously the neurobiologist's version of this story.

Kant had incredible insight in discovering that there are different aspects of our construction of experience that follow different rules. If one of those rules now turns out to be that the accommodation of our neural structures for perceptual aspects of experience is limited to critical periods of development—while we may accommodate to the conceptual aspects throughout adult life—then it becomes a matter of scientific investigation to determine which features of our experience are handled by our faculty of sensibility and which by our faculty of understanding. If modern philosophers are still interested in Kant's original question about how to "apportion" the various features of our experience (and some of them are), they now need to depend upon the a posteriori discoveries of neuroscience as well as the a priori insights of more traditional philosophy.

We saw above how, in the neurobiologist's telling of the story, the faculties of sensibility and understanding are identifiable but have no distinct boundary. There is no distinct point in the processing of visual information where GAP-43 suddenly disappears or where we would want to say "this is the precise point where input processing is done and central processing has begun." It should therefore come as no surprise that the distinctions between a priori and a posteriori and between form and content must both break down when either pair does.

Recall the debate between Hume and Kant as to whether concepts like object or causation were "in us a priori" or "in us a posterior" (Chapter 3). Kant believed the former because he saw these concepts as the necessary form of any experience constructed by our faculty of understanding. Now instead we have a picture of how the "content" of the world—with certain a posteriori constancies found in various features of the way our universe evolved—can write itself into the concepts we apply a priori to determine the form of our experience of that world. And, as these distinctions blend together, we can also remember how Kant's original divide-and-conquer strategy had nature providing the "form" and nurture the "content," so we can be grateful to have found that nature *and* nurture will have to be implicated in both form and content as the boundary between them gives way.

Yet another distinction that has been considered crucial to philosophy but which now begins to disintegrate is that between the "analytic" and the "synthetic." Recall that analytic truths are "true simply by definition" (the predicate is contained in the subject in such sentences as "Joe's father is male.") Synthetic truths, in contrast, contain new information in their predicates (Joe's father is six feet tall). The difficulty with this distinction, once we have entered our circular coherence view of truth, is that what counts as "analytic" depends on our definitions, but these definitions *change* as our science uncovers more successful ways to classify nature.

Distinguished modern logicians and philosophers such as Hilary Putnam and W. V. Quine have dismantled the analytic-synthetic distinction beyond any possible rehabilitation. When scientific definitions no longer point toward any natural uniformity or cleavage, they are abandoned as useless. In the same pragmatic philosophical spirit of his comments above about the nature of logic itself, C. I. Lewis once also wrote that "science is the search for things worth naming." The definition of an "epicycle" before the Copernican revolution or of "phlogiston" before the discovery of oxygen—like the "ether" and many others—were "analytic truths" but revealed themselves not to have any meaning in the natural world. Through the study of nature, the synthetic truths contained therein somehow reveal their influence even on the shape of our analytic definitions. (Perhaps as medicine, science, and society evolve, people will look back and grin at the time "Joe's father is male" was taken to be an analytic truth.)

Before Kant's time, it was in fact generally held that whatever knowledge was a priori must be analytic. "Joe's father is male" could not be proved false through experience, and so it is a priori—but only because it is analytic: being male is part of what is *meant* by being a father. We might now say that there is a (synthetic, a posteriori) *reason* that being male is part of what is meant by being a father, namely, we noticed that nature has always worked that way! If nature worked differently (or if we begin to interfere and change certain biological processes for our own ends—see below), our definitions would be different. So there is a hint of "synthetic" information built into our analytic definitions insofar as they are not sterile and arbitrary but rather products of the times and of the tools we use to help us deal with the times. We might say they are "achievements" resulting from humankind's historical and continuing search through those inadequacies and inconsistencies that reveal broader contexts of the truth (and cause us to abandon the epicycles, phlogiston, and ether). While Kant thus had a classical view of a priori, analytic, necessary concepts that are used to construct any possible experience of the world and are shaped only by our human nature, we can now see the subtle contributions of the a posteriori, synthetic, contingent world to those very concepts. As the multifaceted scholar of the mind Howard Gardner has summarized the entire revolution of the cognitive sciences, "Today it is no exaggeration to say that the *classical* view of concepts has been replaced by a *natural* view of concepts."

This natural view of concepts includes a very dynamic understanding of the relationship between nature and the human beings who interact with it. On the one hand, we might say that, since it is the environment that contributes to the plastic shape of our conceptual apparatus, we should speak of a "*nurtural* view of concepts," inasmuch as the natural world has nurtured our individual nervous systems to accommodate themselves to both the form and content of that world.

On the other hand, we cannot forget our earlier observation that at least some of what is there in the "natural world" is there because human beings put it there. Since our own ideas about the world not only reflect the world but change the world, some other time-honored distinctions become rather blurry as well. Indeed, the distinction between "nature" and "nurture" itself becomes useless when the biological world of nature begins to include new genetically engineered organisms designed to prevent diseases caused by other organisms, and new social institutions are designed to influence the ideas of individuals raised within them.

In fact, even before the scientific revolution that made genetic engineering possible, a broad enough view of nature would lead to questions about any distinction between it and nurture. After all, even the quintessential program of our nature—our DNA—is an end product of the "nurturing" of the environment: the genes that give us a respiratory system that utilizes oxygen were shaped by an oxygen-containing environment. This is perhaps best captured by the phrase, "Heredity is nothing more than stored environment." We have no problem seeing the biological basis of the truth of the statement that "oxygen is *necessary* for our experience," and our analysis has begun to point to the biological basis of similar statements about the experience of spatial, temporal objects as "necessary" for our experience. From this perspective, both the oxygen and the objects are part of nature, and nature has nurtured both our respiratory and our nervous systems to accommodate themselves to the shape of the world. On close inspection, the nature-nurture distinction begins to dissolve. As the psychologist Gregory A. Kimble has said, asking whether our uniquely human experiences of the world are determined by heredity or environment "is like asking whether the areas of rectangles are determined by their height or their width."

Perhaps the most time-honored distinctions of all that begin to break down on this view are the distinctions among those academic disciplines that have been used to try to understand the truth about the world. We had a hint above at how the distinction between ethics and the philosophy of knowledge, for example, becomes blurred when we include in the realm of "things in the world" those products of the human imagination that are there because people thought they should be. If, in our example, even the tables and chairs with which beginning philosophy students become preoccupied have hidden in them—in their very existence—the moral "should" that the founders of their universities built into them, then there can be no clear division between the study of facts and the study of values. After all, values are written into the existence of at least some things in the same subtle way as those things have written their signatures into the concepts we use to experience them. Applying the hard-and-fast distinction between the a priori and the a posteriori that he developed and codified, Kant saw a sharp distinction between the subject matter of philosophy and mathematics (the former) and the subject matter of psychology and the natural sciences (the latter). With the breakdown of this fundamental distinction, we must also accept a new interdependence among these disciplines that is found throughout the pages of this and many other recent books.

Most of the rest of this book will be dedicated to an exploration of how the distinction between ethics (what values should we hold?) and the philosophy of knowledge (what facts should we believe?) becomes blurred within this interdisciplinary perspective, and what the implications of this blurring are for the nature-nurture debate as applied to moral values. It is important, however, to understand how this move from a discussion of knowledge to a discussion of values fits into our coherence wheel of truth. It is simply an instance of what Hegel meant by the "becoming" of knowledge, where he saw that everything short of the *whole* is fragmentary and incapable of existing without contradiction unless complemented by the rest of the world. The hidden "should" we discovered in those tables and chairs is an excellent example of what Russell called Hegel's "metaphysical hooks" that grapple one incomplete conception of the world to the next as we strive to achieve a more complete coherence in our understanding of the world.

Our task thus becomes the examination of each of our successive conceptions of reality to discover what internal contradictions might lead us to a more adequate understanding of the world and so "uncover in our thoughts the signatures of the helping hand of things." Following this Hegelian project, we carefully search the many close-knit relationships between thoughts and things only to discover that some of the things about which we have thoughts have yet another important feature: that they are only *there* because of values (thoughts) held by human beings (not only contemporary human beings but also our ancestors), and this becomes a fresh hook to use in the next step of our reconstruction of reality.

But before moving on to a more detailed discussion of ethics, we need to take a look at what remains once all of these old distinctions have broken down. Once we no longer have clear boundaries between a priori and a posteriori, between analytic and synthetic, between the form and content of experience, and even between the disciplines we use to study ourselves and our nature (and indeed even between nature and nurture themselves), what can replace these distinctions so long considered crucial to our thinking?

Part Three

Men are neither angels nor devils; that makes morality
both necessary and possible.

H. L. A. Hart

15

Everything Is Relative?

Of all the traditionally important distinctions that become blurred when we adopt an interdisciplinary perspective, perhaps the most significant is the time-honored distinction between subjectivity and objectivity. This is a much-abused distinction, but for definitional purposes we might remember that Kant sought the status of objectivity for his knowledge about the world by grounding it in a priori concepts that must necessarily be applied in the construction of any possible experience. The "objective" in this case is taken in a very strong sense, since "necessarily" referred to the absolute requirement that any sentient being capable of having any possible experience would "necessarily" apply Kant's conceptual categories (including Martians were they to have been found to exist and have experience).

We might say that the "reference group" Kant intended to implicate in his claims about "objective knowledge" was any being capable of experience (human or otherwise). This is presumably the largest frame of reference a theory of knowledge could demand. In the strong, Kantian sense of objectivity, my knowledge of the world as spatial, temporal, and containing permanent objects makes a claim on every other knower's beliefs about the world. Since we all necessarily apply these categories to construct any possible experience of the world, I am entitled to tell anyone who claims the world does not have these features that they are either kidding around or just plain wrong. At the opposite extreme, the term "subjectivity" implies a frame of reference of only one: I make no demands on another person's beliefs when I claim to have the subjective experience of enjoying an ice cream sundae or having a tickling sensation in the back of my throat.

Between these two extremes fall the many intermediate-sized reference groups involved in the demands we daily make on other people's beliefs about, say, art or music. I share with *some* other people a belief that good musical comedy is one of the highest forms of art, but I do not share this belief with *all* other people (not even *most* other people). On the other hand, within this somewhat eccentric reference group, I have been convinced by fellow group members that I was wrong and they were right about whether a particular specimen by Rodgers and Hammerstein or Gilbert and Sullivan makes the cut to count as good musical comedy or not. Standards of evidence emerge for demands on knowledge within such intermediate-sized reference groups, even if we members of the group do not hold most other (more musically sophisticated) people to these standards.

As we have begun to view human experience not as possessed by some theoretical knowers but as it arises in actual biological beings such as ourselves, we have come to a very different view from Kant of the mental structures we use to construct this experience. This is in part because we have availed ourselves of empirical research on the genesis of these structures—whether through Piagetian-type research on basic concept acquisition, psychoanalytic observations of emotional development, or neurobiological investigations of brain development.

Exactly what size frame of reference can be demanded for the claims of a human knowledge that arises through the nature-nurture interactions we have discussed? Kant made some serious assumptions about what must "necessarily" be true for all knowing beings, but we have now seen how the development of actual human knowledge depends upon such conditional factors as adequate nutrition in utero and good-enough mothering in the first years of life. If Piaget subsumed the interaction between thoughts and things as part of a subset of all interactions between organism and environment, whatever could bind all people into one big reference group in this biological system would have to depend upon certain "constancies" reliably appearing on *both* sides of the nature-nurture interaction: the *organism* and the *environment*. The mixture of genes and environment that leads to a love of musical comedy cannot be counted on to appear on both sides (or even either side!) of the interaction in the case of most people, and so we make no demands on their beliefs about this underrated art form. But there is an irresistible urge in the modern world to demand that *all* people believe that the earth goes around the sun and that objects are permanent, or else we say that they are either psychotic or just plain wrong. What could possibly be so constant in the nurturing environment and in the biological nature of all these billions of organisms that our frame of reference for most knowledge about the world is presumed to be all human beings?

As for the environment, our biological version of this story is easy. What we mean is simply the physical world as biology describes it. As we noted above, it is no small matter that this world is everywhere and unavoidable, for these are properties which guarantee reliability on the environment side, both within and between individuals. This does not only mean that the everywhere and unavoidable world nature harnesses in order to nurture the developing brain must be manifesting the ubiquitous conditions that make *any* life possible (as in an "anthropic principle" notion that time, space, permanent objects, and the like may be necessary features of any universe that could produce living beings); it also means that these constancies will reliably appear in the environment of *each individual* knower. As we saw in our discussion of the facts of life (Chapter 9), nature is not taking much of a risk in harnessing itself to nurture with reflexes like orienting the infant's eyes to human faces to help orchestrate the wiring of our cognitive pathways, since any infant that will survive to have knowledge of the world will have had *some* other person's face around during the prolonged period of dependency nature has crafted for us. The same goes for the intrauterine environment during embryogenesis. As our review of the facts of life also revealed, human intrauterine environments have much more in common with one another than they have differences. When a given uterine environment is significantly different from nature's prescription, fetal development does not proceed.

As for the organism, our biological version of this story is equally straightforward. The individual cognitive structures that are developing within each person, once understood biologically, arise from a complex interaction of the individual's genetic endowment and the internal and external environments in which the central nervous system develops. There are at least some constancies that we can depend upon if human life and human experience will become possible. Even Piaget, with his detailed attention to individual children, believed himself to be concerned with those cognitive structures which "if they hold true for the individual, also hold true for the species." Piaget did not deny that individual differences, motivation, and so on may affect the acquisition of these structures, but he insisted that such individual variations cannot affect the identification of these structures, which as products of evolution he viewed as *species-defined* structures.

At least some basic rules for brain organization are certainly genetically programmed, but as we saw above, we need not rely on identical programming for sufficient constancy on the organism side to guarantee that species-defined cognitive structures will be produced in all individuals capable of human experience. Although the fine detail of the individual wiring in different people's brains is inevitably different, the mechanism nature devised to give rise to those connections was sufficiently ingenious to ensure that certain everywhere-and-unavoidable features of the world will write themselves into that wiring. In our discussion of the theory of neuronal group selection we thus saw Edelman insisting that selective processes guarantee both uniqueness of fine neural wiring and certain common patterns across the species. We might now use our new language and suggest that the frame of reference for at least some of our concepts is the entire species when these concepts—these common *patterns* in neural wiring—are contributed by the most pervasive features (object, number, causation) of our everywhere-and-unavoidable world.

But does this species-wide reference group make these concepts "objective" truths about the world in Kant's strong sense that they are "necessarily" features of any experience any being could ever have of the world? Although we are tempted to say that all human beings may be bound by the claims of these truths, the many contingencies built into our biological version of the story—from the child's surviving intact to evolution's having gone the way it happened to go—should make us circumspect about staking out the philosophical status of "objectivity" for these truths. But what other word will capture this status?

On purely philosophical grounds, Wittgenstein (whose sandy riverbed metaphor above so perfectly captured Hegel's more abstruse "becoming of knowledge") attacked the notion that even species-binding beliefs about the world could be defined as "objective" in any way that meaningfully distinguishes this from "subjective" experience. As mentioned above, Wittgenstein could have easily launched our entire project (instead of Kant and Hegel) if we were working within the idiom of modern linguistic philosophy, since his insights led philosophers for the last half century to think of our "categories of understanding" in terms of the rules that govern language.

Our interdisciplinary approach resonates with Wittgenstein's philosophical attack on the distinction between the subjective and the objective in much the same way that it resonates with the attacks by Putnam and Quine on the distinction between the analytic and the synthetic. In this case, what makes the traditional distinction between the subjective and objective especially problematic was our discovery that the actual environment in which each individual subjectivity approaches the objective, external world is *necessarily another subjectivity.* Winnicott insisted there is "never just an infant" but only an infant-mother dyad, since any developing subjectivity necessarily comes "attached" to another. The "necessarily" here is slightly different from the "necessarily everywhere and unavoidable" description of our biological world, but it is also firmly grounded in biology. Given that crucial biological fact of our prolonged dependency upon adult members of the species for survival, we might conclude from this perspective that neither is there such a thing as a "subjectivity" but only what we called in Chapter 7 the intersubjectivity of the collective experience of that dyad. The expression *intersubjectivity* tries to capture the social aspect of even individual subjective experience. In Wittgenstein's linguistic terms, even the concepts that shape my most personal (private) thoughts must be framed in terms of a language that has to be *inter*personal (social) if it is to have any meaning. We might therefore use the term "intersubjective" for the status of truths that—while binding on all people—are not necessary in the strong sense Kant intended for his objective truths. What can we say about these intersubjective truths?

If, in shaping our plastic brains, intersubjective contributions from other people are necessarily made in the formation of mental structures in any human destined to survive and become capable of having knowledge, it may be possible that some generalizations can be discovered about even their highly individualized mental or neural structures. The reason this approach to the question of objectivity versus subjectivity demands an interdisciplinary perspective is that some of this generalizing work requires a view of the human being from the outside—of the human being as a phenomenon in nature, subject to the laws of other natural phenomena. This was the case in our discussion of the neuroscientist's view of the matter, where we do not posit a conceptual mind struggling to understand the laws of nature, but a biological mind which is itself a part of nature and subject to those same laws. In this view, Hegel's mystical-sounding dictum that "reason is relative to reality" becomes common sense if reason is an achievement of our minds and we take our minds to be the actual products of the biological world. The famous slogan "What is rational is actual and what is actual is rational" reminds us that we have obtained our notions about reason from our observations of and participation in the actual world. As we saw above, even our analytic definitions—and even the basic principles of logic—are ultimately supported on pragmatic grounds. If, like philosophers in the Cartesian tradition, we try to abstract ourselves from the world and view the mind as a private thought-theater independent from the world, we will understand neither ourselves nor the world. After all, as Wittgenstein highlighted, whatever comprehension is possible of any "private self-conscious subjective" experience depends upon an extensive backdrop of "public" practices, rituals, relationships, emotions, and not least, *language:* a backdrop which gives meaning to the concepts that we apply in constructing our experience in the first place.

Our discussion thus far has de-emphasized the role of language (for reasons discussed above), but it is impossible to overestimate the importance of words once we enter the interpersonal environment. Words are used both as a means of communication and also as one type of symbol available to the brain to increase its efficiency dramatically in absorbing information from the world. As soon as language is introduced, however, questions may be raised about all of our facile assumptions above concerning intersubjective limits to any possible human experience and the "constancies" that support it. Philosophers throughout the centuries have been intrigued, for example, by whether it is possible that when you and I each use the words "blue" and "red" our inner experiences are actually reversed for these colors. This possibility is potentially fatal to a philosophy of knowledge that views as its standard a coherence that is intersubjective—that builds successively broader understandings of the world by comparing each not only with our own past understandings but also with those of our fellow thinkers. After all, the question goes, if I were born always *subjectively experiencing* as red what you *subjectively experience* as blue, would not we both learn to use the same words in all situations while intersubjective coherence would hardly hold? Intersubjective coherence demands a frame of reference of the species: the challenge is raised here whether true knowledge of something as basic as the redness of an object might include a reference group of only one individual—a pure "subjectivity."

Before we address this either-or question, we might wonder whether claims about colors have some intermediate-sized reference group based on specific characteristics of groups of people greater than one but smaller than the species. For example, is it possible that "redness" is not purely subjective and not truly intersubjective but relative to all English speakers? The idea here would be that the word used for "red" in another language might be coherent within that culture but refer to an experience of light reflectance we English speakers see as blue. This would be a quintessential example of Whorf's hypothesis that culture (here, language) determines experience.

Color is an excellent example because we know from both aesthetic experience and science that colors exist on a smooth continuum of the rainbow: wavelength is a continuous function and it is we humans who, in experiencing the reflectance of lightwaves through a mechanism involving three particular kinds of cones in our retinas, apply discontinuous labels like "red" or "blue."

Indeed, the classical view of color experience was that the appropriate reference group was a given culture, with the words used for "red" or "blue" defining how that culture's experience divides up the spectrum. In fact, this classical view received some support when cross-cultural studies focusing on the border regions between colors found some differences between cultures in precisely where they draw the line between "red and orange" or between "blue and green" measured "objectively" by wavelengths. This early work implied a certain arbitrariness to color naming, with various reference groups (in this case, cultures) differing in their experience of the color universe. This was taken as strong support for Whorf's ideas, and people actually began to wonder whether, because Navajo collapses blue and green into one word, members of the Navajo Nation do not experience a difference between blue and green (like Hopis not distinguishing past, present, and future)!

In the 1970s the cognitive anthropologist Eleanor Rosch and her colleagues traveled the world proving that the color universe is not so arbitrary or linguistically defined as Whorf's view implied. Rather than focusing on the boundaries between colors, Rosch used a spectrum of colored chips to investigate what various cultures would consider a "good red" or "pure red." The universality—in every language and culture—of where on the spectrum one would find the prototypical "red" was a serious blow to Whorf's hypothesis, supporting instead the language-instinct theory that we are biologically conditioned to see colors the way we do, and all languages ultimately apply some words to represent this (intersubjective) experience. As

Pinker wonderfully describes the outcome of this discovery, "Although languages may disagree about the wrappers in the 64-crayon box—the burnt umbers, the turquoises, the fuchsias—they agree much more on the wrappers in the eight-crayon box—the fire-engine reds, grass greens, lemon yellows." But this anthropological finding was just the start.

Indeed, Rosch's predecessors in this work, Brent Berlin and Paul Kay, had previously discovered a universality to the order in which primitive cultures come to assign color names, with very primitive cultures only having words for black and white. More advanced cultures sequentially add red, then yellow and green (in either order), then blue, then brown, and finally pink, purple, orange, and grey (these last four in any order).

What was left for Rosch and her colleagues was to find a very primitive culture to study. This she found in the Dani of New Guinea, a stone-age culture that has but two color terms, "*mola*" for bright or white and "*mili*" for dark or black.

On testing with colored chips of various hues, Rosch found widespread agreement among the Dani at extremes of *mola* and *mili* but found considerable individual variation on where subjects located the boundary between the two. Instead of accepting this as evidence for extreme subjectivity, Rosch performed an ingenious experiment. Using Dani names unrelated to colors, Rosch created a vocabulary for seven colors. However, half the Dani were given a vocabulary which identified those wavelengths defined in more advanced cultures as "prototypical" examples of each. The other half were given the same vocabulary, but applied to "intermediate" wavelengths.

Unexpectedly, the Dani given words for prototypical colors learned the vocabulary quickly and remembered it easily—and in fact recapitulated the evolutionary order of color naming!—while the group given words for intermediate wavelengths seemed unable to learn or retain the new vocabulary. There is, it appears, a *natural* way to divide the spectrum into color categories, a division which appears to have more to do with color itself and our human mechanism for experiencing it (our three types of cones and how they are wired to our neurons) than it has to do with any particular culture or language. Be reassured: members of the Navajo Nation experience blue and green just like everyone else who is not color-blind (more on that later). While the classical view maintained that our categories are arbitrary (depending on language or culture) and have sharp boundaries, this research implies that there are *natural* ways to categorize even continuous aspects of the world, and these natural categories have blurred boundaries.

These biological facts of our cognitive life are completely consistent with Wittgenstein's insight that our language categories all have such prototypical "centers" (a "good red," a "good chair") but fuzzy boundaries (reddish-orange, a piano stool). Wittgenstein not only recognized the interpersonal, social element in calling something "red" or "a chair"—which is the crux of his famous argument against the possibility of a "private language"—he also recognized that membership in any category works like membership in a family. This natural human way to categorize leaves us with an experience of the world where family resemblances ensure that all members of a category bear some similarity to the prototypical center, but more distant relatives even within the same category can actually have surprisingly little in common. (My house is a home and so is Buckingham Palace.)

When Gardner summarized how the classical view of concepts has been replaced by a "natural" view of concepts, he did not yet know how Edelman's discoveries would offer a neurobiological grounding for the way our brains generate such concepts. In an interesting philosophical analysis, Edelman has demonstrated how the concepts generated by neuronal group selection operate according to Wittgenstein's idea of family resemblances. Since neuronal stabilization with repeated functioning is just a statistical phenomenon, members of conceptual categories must have just such degrees of membership with central prototypical members and blurred boundaries. This is just the "natural" way categories are formed by our brains.

To talk of "natural" categories is not to contrast natural with man-made, since we continue to take humans as *part of nature,* not as unnatural subjects observing nature. Thus, when work similar to the color study investigated the categorization of various geometric shapes, the results demonstrated that circles, squares, triangles, cubes, and the like appear to be the natural shape categories for people to learn and remember even though these shapes do not exist in "nature."

Given our analysis—based in biology—of the interactions of organism and environment that include such constancies as preferences to look at faces and now even how the visual system divides up the continuous color spectrum, we can better understand how intersubjectivity can offer an alternative to radical subjectivity (with its inability to make any claims at all about knowledge between people) and objectivity (with its demand for "necessary truths"). The insightful contemporary philosopher David Wiggins has offered one suggestion to those who would try to force us to choose between the subjectivist view that "postboxes are called 'red' *because* we see them that way" and the objectivist view that "we see them that way *because* they are red."

> Maybe it is the beginning of real wisdom to see that we may side against both . . . and ask: "why should the *because* not hold both ways round?" Surely an adequate account of these matters will have to treat psychological states and their objects as equal and reciprocal partners, and it is likely to need to see the identifications of the states and of the properties under which the states subsume their objects as interdependent (if these interdependencies are fatal to the distinction of inner and outer we are . . . to be grateful for that).
>
> We may see a postbox as red because it is red but also postboxes, painted as they are, count as red only because there actually exists a perceptual apparatus (e.g. our own) which discriminates and learns on the direct basis of experience to group together all and only the *de facto* red things.

Although Wiggins bases his entire argument in philosophy and psychology, we can add our knowledge of neurobiology and of the organization of both the external world of things (objects) and the inner organization of human minds (subjects) to see how it is impossible for you and me to come to *know* redness and blueness in opposite ways. Of course, we may occasionally perceive the same things differently (especially if conditions are purposefully set for illusions), but we would not say that we had *knowledge* that failed our test of intersubjective coherence. Remember our experience with those Müller-Lyer arrows way back in Chapter 3. Although some of us (but not all, as we shall see later) *perceive* the Müller-Lyer arrows as unequal, we still *know* them to be equal because we base our knowledge claims on an intersubjective (universal) coherence that takes advantage of the "metaphysical hooks" discussed above—the way we search our knowledge for the inconsistencies and inadequacies that make perceptual differences epistemologically benign. We *all* perceive the earth to be stationary with the sun circling us instead of vice versa. If some primitive cultures still believe that the earth is not moving, our own further movement along Hegel's becoming-of-knowledge allows us to make a claim against the *truth* of this belief.

Many writers have contrasted the "objective" with the "relative," and there is indeed one way in which our biologically based intersubjectivity is still relative (which is why it is not "objectivity") but not, as we have seen, culturally relative. We might say that intersubjective truth is relative to the biological facts of nature. As Wiggins put it in his argument:

> Not every sentient animal which sees a red postbox sees it as red. Few or none of them do. But this in no way impugns the idea that redness is an external monodic property of a postbox. "Red postbox" is not short for "red to human beings postbox." . . . All the same, it is in one interesting sense a *relative* property where the property of color is an anthropocentric category. The category corresponds to an interest which can only take root in creatures with something approaching our own sensory apparatus. [emphasis in original]

Put another way, what intersubjective truth is *relative to* is our actual biological world containing human beings such as ourselves. Reminiscent of the anthropic principle—where even the laws of physics are in some ways relative to human knowers—this type of relativism is no loss, and it is in fact potentially quite a gain. Intersubjective truth is considerably less than true "objectivity" (which implies some *necessity* in a very strong sense), and we instead must accept that evolution is still in progress and we in some ways only "happen" to be the creatures we are at this moment in the universe. But that strong "necessity" of objective truth has often been considered by modern philosophers to be the most objectionable aspect of it, so this is not much of a loss.

Indeed, the main advantage of objectivity has presumably been its *universality*, since pure subjectivity insists that empirical truth is not only contingently true (it might have been otherwise) but is also only *locally* true (true for some subset of beings capable of knowledge). Intersubjective truth is universal, since coherence is required at the level of all human knowers, but it is contingent, since it was shaped by evolution going the way it has. When a mystic might claim that he sees red and blue reversed subjectively, or a psychiatric patient claims that objects are not permanent and do not persist through time, we can therefore understand and appreciate what they are trying to communicate, but we can ultimately insist that the claim about *knowledge* they are making is false.

The intersubjective position advances the possibility of *universal contingent truth*, thus providing relief from the *necessity* of objectivity and the *locality* of subjectivity. From this position, even reason itself becomes a "contingent feature of our changing world." In blurring the distinction between the contingent a posteriori facts *of* nature and our own a priori concepts *about* nature (through the accommodation of the former into the latter), we have completely abandoned the "purity" of Kant's view of reason and discovered, in a phrase, the *contingency of truth*. The traditional philosophical demand that the truth be "objectively necessary" now almost becomes hard to define in any meaningful sense, since the very words we use are intimately bound up with the real world.

In Chapter 5, we noted that there is a certain amount of historicism inherent in any theory based in biology and therefore evolution. For this reason, even Edelman the biological scientist does not call himself a "realist" philosophically, but rather a "qualified realist," since "our description of the world is qualified by the way in which our concepts arise." When Hegel the philosopher likewise reminds us that even reason itself is "relative to reality," and the psychoanalyst Winnicott reminds us that there is "never just a subjectivity," we can take this as a new kind of interdisciplinary coherence about the intersubjective status of even our most "indubitable, incorrigible, universal, and absolute" truths.

16

Inventing Right and Wrong

Piaget reminded us that whatever relationships hold between thoughts in our head and things in the world must have some characteristics in common with the greater set of rules that pertain between the organism and the environment. On this view, *thinking* is just one example of the many ways we adapt to our environment, and so thinking must have common features with taking in nutrition, defending ourselves from predators, and other ways all organisms adapt to their surroundings. As we move away from the philosophy of knowledge to consider the field of ethics, what we shall find is that virtually all of the general arguments that were made about the relationships between thoughts in our head and things in the world apply equally to the relationships between "moral values in our head" and "social practices in the world."

But rather than begin with these relationships and proceed in the same order of argument, let us first recapitulate the conclusions found in the previous chapter for the cognitive realm of facts to see where such reasoning will lead us in the moral realm of values. This will be instructive because the presuppositions of the modern secular world tend to assign our adherence to facts and our adherence to values at opposite ends of the objectivity-subjectivity continuum. Despite the strong attack that modern philosophers have begun to launch against Kantian cognitive objectivism over the past thirty or so years, most people still have quite an emotional attachment to the idea that "true facts" necessarily *have* to be true in some strong Kantian sense. This makes cognitive intersubjectivity something of a disappointment, since facts (including scientific facts) are no longer "objective" in this stronger sense. The anthropic principle is counterintuitive because most people would like to believe that facts are not *in any way* relative to those who know or discover them. In this popular view, facts are thought to be built into the fabric of the world and are *discovered*, not *created* by society.

In sharp contrast, there is a strong tendency in the secular world to deny the existence of anything like objective values. Values, after all, are meant to tell us not only about how the world *is* but also how the world *ought to be*. A claim to objectivity in the moral sphere would require that our objective values are also somehow built into the fabric of our world. They would have to already be there to be discovered through some effort of the will, through some rational exploration or some intuitive leap of understanding. In the secular world it is difficult to see how truths about how the world *ought to be* (in the future) could be built into the way the world *is* (in the present) and so be available for discovery. Evolution at best selects for states of affairs that *have been* adaptive (to past circumstances). "All values are relative" becomes the slogan of the day, uncritically accepted by those who have come to see this position as the only viable alternative to the notion of necessary, universally binding standards that "objective values" would represent. Cultural imperialism and religious warfare are the images that come to mind when most people contemplate the implications of moral objectivism in the modern world.

Despite the advances made recently in the academic halls of philosophy, the marriage of moral relativism and cognitive objectivism comes naturally in today's world. These positions are built into our modern language and ordinary (commonsense) thinking. Yet they represent an incoherent combination of positions. There are three reasons for this.

The first and perhaps simplest reason that facts and values must stand or fall together on the question of objectivity versus relativism is that there is no clear boundary between them. We have already seen many of our time-honored distinctions blurred, and the fact-value distinction is no exception. The continuum on which both facts and values reside is highlighted by the existence of "social facts" which fall between the two, as the philosophers Steven Lukes and William Runciman have demonstrated with the following example. When we ask the "factual" question "Who has the power in society?" we are clearly talking about something which exists somewhere in the middle zone between factual statements and value statements, as one person answers "the current political regime," another answers "the intelligence agencies," another "the multi-national corporations," another "the church," another "the working class, though they themselves do not yet realize it," and so on. With no boundary separating the two, it is hard to see how facts and values could reside at the opposite extremes of such a fundamental philosophical divide.

A second reason that moral relativism and cognitive objectivism represent an incoherent combination of positions is that we have already seen how cognitive objectivism is itself incoherent: there is even in the cognitive realm a subtle relativism to the natural world as it historically developed that is intimately interconnected with any facts we can know. The presumed distinction that we encounter nature and *describe it* to achieve facts but create social practices and *prescribe* values was dealt a counterintuitive blow by the anthropic principle, as cosmologists have now discovered the extent to which we "prescribe" even the basic laws of nature by being here to ask questions about them (again, not by causing those laws but by constraining them to possible laws that could have caused us). And at an even more practical level, we saw how so many of the objects we encounter have hidden moral values built into them, since they only exist because people thought they *should* exist.

Finally, a third reason to question how facts and values can attain such radically different statuses is that they arise out of the same brain mechanisms. We have already seen how values and perceptions are both necessary for any categorization at all to arise in the brain (Chapter 12). This more than anything else should make us wonder how our concepts about what deserves naming (in science) and what deserves valuing (in ethics) could end up on opposite sides of the great philosophical divide. Arising out of the same neural processes as we interact with the world around us, our facts and our values must ultimately stand or fall together on the question of objectivity versus relativism.

Of course, our cognitive theory ultimately found the truth of our facts at neither of these extremes but in the middle position we called intersubjectivity. In ethics, the intersubjective position arises with the following challenge to our modern relativistic assumptions: "Well, if all values are relative, exactly what is it that they are *relative to?*" The answer typically comes back that values are relative to the needs, wants, and interests of the human beings appealing to them. Since these needs, wants, and interests usually arise from characteristics shared by more than one person, we quickly generate a host of overlapping spheres of groups to which values are "relative." Very often, the appeal is made to the needs, wants, and interests generated by characteristics peculiar to a given *culture.* We would then say that these values are "relative to" that culture, and we would join the popular position called cultural relativism. Here, the value attached in a small African tribe to bowing three times upon walking by the chief's son is taken as a culturally relative value.

But cultures define only a subset of the possible reference groups that share relevant human characteristics. Because of common needs, wants, and interests, some values are likely shared by all urban dwellers, regardless of culture, while others may be shared by all rural dwellers. Similarly, some values are likely shared by (relative to) any minority group within any culture, be it blacks in a predominantly white culture or Moslems in a predominantly Hindu culture. Perhaps some values are even shared by each of the sexes (women as a group may have some needs, wants, or interests not shared by men as a group). And similarly for all groups, if by "group" we mean—as we always do—a collection of people with some shared identifying characteristics.

But human beings also form a group unto themselves, and there are some needs, wants, and interests that are generated by the identifying characteristics defined by the term *Homo sapiens.* These are largely biologically determined characteristics, so our species-defined needs, wants, and interests are also some of our most basic and important ones. *All* people require food, shelter, clothing and so on. Some, myself included, would put on this list such characteristically human requirements as love, attention, and opportunities for self-expression. To call these needs, wants, and interests species-defined is not to say that they are species-specific. Many animals need shelter, and virtually all animals need some kind of food. But our earlier review of the facts of life did highlight the biologist's version of the necessary conditions for any possible human experience, and these included whatever relevant characteristics arise out of not only nine months of gestation but also the prolonged period of dependency nature has created for us and whatever else may be needed to continue to survive into adulthood. We presumably share some of these characteristics with our closest animal relatives, but for now we will limit ourselves to a consideration of human beings.

The creation of a complete list of such species-defined human needs, wants, and interests is not so important here. (Those interested in one rather complete version are recommended to anthropologist Donald E. Brown's recent book *Human Universals.*) What is important for our present purposes is an appreciation that the values generated by these defining characteristics of our species will be "relative to" *every human being in the world.* Such values will not be "objective" in Kant's sense of "necessary" and "built into the fabric of the world"—indeed, if there were no humans, this particular set of values would not exist at all. While they are thus contingent upon the way the world happens to be, these values are also *universal,* since they apply to every human being who ever lived or ever will. These values may therefore be called *intersubjective values* in exactly the same way we have called some facts *intersubjective facts.*

In considering the many overlapping spheres of shared human characteristics which may give rise to shared needs, wants, and interests (be they cultures, genders, or the whole species), we may again speak of the reference group of values as that group whose shared characteristics give rise to them. Intersubjective values may well be contingent (in that they presumably could have been otherwise), but their frame of reference is *all human beings*. Like their cognitive counterparts, intersubjective values (as well as facts) are "universal contingent truths."

It is instructive to compare values with smaller and larger reference groups. The brilliant British philosopher John Mackie has pointed out (using a very different language) that values with the largest frame of reference—truly intersubjective values—tend to be the most general in behavioral content and the most stable over time. We still leave open many different ethical ways to live in the world when we make the intersubjective moral claim that it is wrong to torture and kill people just for fun, or that it is better if people have some opportunity for self-expression. The shared human characteristics that support these claims could only change on an evolutionary time scale. In contrast, values with smaller reference groups tend to be the most specific in behavioral content, but they are also the most changeable over time. Bowing three times upon walking by the chief's son is very specific, but we are not surprised that it becomes a simple salute a generation later or even disappears a generation after that. Intersubjective values (universal contingent truths) may always be quite general in behavioral content, but we would be surprised if they changed very much in a generation or two.

Despite their generality, intersubjective moral truths do provide some concrete guidance in certain important spheres of human activity. For example, intersubjectivity offers an answer to the cultural relativist who would make the claim that when a foreign government tortures innocent people, we in our country cannot make any claims about the morality or immorality of such acts. It is in fact biological characteristics shared by *all* human beings which inform us of the immorality of torturing innocent people, and so we *are* entitled to condemn such acts as immoral. Intersubjectivity can therefore provide some ethical basis upon which to ground decisions on international (intercultural) policy. If what we mean by basic human rights are these intersubjective values which are generated by the shared characteristics of all humanity, then this thesis would support (indeed would insist upon) an appeal to these basic human rights (intersubjective values) in providing moral justification for foreign policy decisions.

In the chapters that follow, we shall explore some of the implications of this line of reasoning for matters of practical concern to our political economy. But let us first elaborate in more detail on the parallels between our earlier cognitive arguments and our forthcoming moral arguments. We have seen how modern prejudices would insist on a radical difference between the way we treat the cognitive dialectic between ourselves (with our thoughts) and the world (with its things and its physical laws) and the way we treat the moral dialectic between ourselves (with our values) and the world (with its social practices and institutions). But both arise from the same interaction of real neural processes with the real biological world. In the former case, this cognitive dialectic was eventually described in terms of two equally valid "becauses." As Wiggins explained, we can say both that postboxes are called "red" *because* we see them that way, *and* that we see them that way *because* they are red. The truth of both came out of our synthetic understanding of the ways we both assimilate the world and accommodate to it in the course of coming to "know things."

We can now see this same dialectic in another two "becauses" having to do with values. We consider opportunity for self-expression to be good *because* we humans need and desire it, *and* we humans need and desire opportunity for self-expression *because* it is good! As Wiggins said, it is the beginning of wisdom to see that we do not need to choose between Spinoza's "cognitivist" view that "it only seems good to us because we desire it" and Aristotle's "noncognitivist" view that "we only desire it because it seems good." Although in our over-intellectualized way we are always tempted to separate our assimilation of the world from our accommodation to it, we must remember that Piaget's original coining of these terms came out of a practical biological view of how organisms adapt to their environment. One scientist may focus on the ways in which the food (or the lightwaves coming from a tree, or our complex social practices) is cut, chewed, and digested in the process of assimilating it. Another scientist may focus on the ways in which the organism changes the shape of its mouth, throat muscles, and digestive juices (or neural structures, or value system) in the process of accommodating to it. But in the real world all we are talking about is the unitary act of having a meal (seeing a tree, engaging in moral life). When, in the practical world of biology, all that is happening is someone is eating some food, we must acknowledge that it is only *we*, in our overly cognitive enthusiasm, who choose to separate the perspective of the food being chewed from the perspective of the person doing the chewing. The two "becauses" are likewise only separable as perspectives in the intellectual sphere and are, in fact, part of a unitary process we call living in the world, that same larger context that subsumed "knowing" above, and now subsumes "valuing" as well.

Based as they are in our naturalistic philosophy (that grounds even our most abstract concepts in the realities of our biological life and social practices), intersubjective values tempt us to think of them as some kind of natural moral law. Indeed, there was a hint of this in Edelman's insistence that even our most sophisticated moral distinctions must have some connection to the deeper value structures that nature has provided (and that underlie our capacity to make any distinctions whatsoever). But *natural* must now be taken in the most literal sense as deriving from *nature*, from the biological facts of our actual human world. Just as *reason* is "relative to reality," so is *morality* "relative to reality." To the extent that we have determined that there is an inextinguishable contingency to even our basic truths, the theory of intersubjectivity is quite unlike most theories based on natural law in that we cannot take for granted our ability simply to discover values in the fabric of nature. If we concluded that even factual knowledge "becomes"—that is, grows and improves with the times—then values likewise are not static. If *science* is the "search for things worth naming," then *ethics* is the "search for things worth valuing." Unlike the "natural values" that grow out of traditional theories based on natural law, which in modern times would make their direct appeal to nature understood as the process of evolution (and which could at best be looked to in support of characteristics which *were* useful in adapting to earlier environments), our intersubjective values must also grow, must also *become* as we search our experience for those subtle inconsistencies and inadequacies that will hook us to ever-broader contexts of living. To the extent that they are natural moral laws, they are not so "self-evident" as Thomas Jefferson believed, but are themselves *achievements* as we search our values like we search our concepts for the inconsistencies and inadequacies that reveal larger truths. Thus, to use one of Mackie's examples, if patriotism is a value which was once useful, but is no longer, then the "becoming of knowledge" can open our minds to take in this larger context and help us come to know ever *better* truths and ever *better* values.

But before we can attempt to identify the process by which we might begin this search for better values (and elaborate on the meaning of this counterintuitive phrase), let us look briefly at the larger institutions within which the moral drama is enacted. Let us look at how morality "works."

17

The Prisoner's Dilemma

The past two hundred and fifty years of moral philosophy have raised some questions about whether one can move so freely as we have here from factual statements about the world (for example, descriptive statements about the nature of our species) to ethical statements about how humans ought to behave—Hume's famous "leap" from "is" to "ought." Mackie has reminded us, however, that the movement from description to evaluation is clearly justifiable within some given institution. I can tell my friend that it is "wrong" to move his rook diagonally when we are playing chess because such a move is not allowed within that enjoyable "institution." Mackie reminds us also that "institution" can refer to very broad classes of behavior. We may, for example, move from the factual description of a promise being made to the evaluative conclusion that the promiser ought to try to keep the promise because the people involved have entered into the "institution of promising" by uttering and accepting the words and actions they used in the course of promising. So not all institutions have to be "instituted" (like the game of chess), but may instead refer to such universal human practices as promising.

It is in this broad sense that the philosopher G. J. Warnock discusses morality as itself an institution—an institution which exists for a purpose, namely, to make things turn out better for all of us in those ways in which they would likely go quite badly if we all always acted on our own selfish impulses. This notion that there is a *point*, an *object*, to this institution we call morality may be understood as relating to the real world in the same practical way that our knowledge relates to the real world. What we mean by things going "badly" without the values of promise-keeping, truth-telling, and so forth is that things will go badly in the natural (nonmoral) sense that human needs, wants, and interests will be frustrated. Once morality is understood as an institution—an institution whose aim is to ameliorate some of the limitations inherent in the human condition—then our values derive their meaning within the living context of the biological world (with its inherent limitations), and we may be justified in moving from "is" to "ought." This view of, say, the practice of promise-keeping or truth-telling is consistent with Piaget's view of all such practices (knowing, eating, valuing) as mechanisms of adaptation to the environment, where we now take the environment to include the *social* environment. Indeed, Lawrence Kohlberg, who spent over twenty years studying the ways in which children accommodate to the values embodied in the myriad of social practices we call "morality," proposed a series of Piagetian stages through which children accommodate basic values of promise-keeping, truth-telling, and so on (recall the chart on page 71).

Perhaps the most famous proponent of this view of "morality as a device" was the influential seventeenth-century British political philosopher Thomas Hobbes. Hobbes saw the human condition as worse than a mere "predicament": he saw life in the premoral state of nature as "solitary, poor, nasty, brutish, and short"! For Hobbes, "short" really was the clincher, since he believed that without some such collective device, humans in the state of nature would quickly kill one another off. He appreciated that there were limited resources for which people naturally compete, but he was more concerned with the limitations on people's sympathies for one another than with the limitations of available food, shelter, and the like. Hobbes's *Leviathan* (1651) is a monument to the belief that our moral, economic, and political institutions can counteract these natural limitations on our sympathies for one another. He believed that it would be in each person's self-interest to give to the state absolute authority, so long as every other person did the same. That way, although we are no longer free to kill one another, we are also now free *not* to be killed by one another, since the state will enforce rules, punishing anyone who acts on this "premoral freedom" to kill that becomes abolished in the moral society established under this all-powerful political authority.

It is interesting how many elaborate rules of ethics (keeping your promises and the like) Hobbes attempts to deduce from his basic assumption that survival-is-better-than-not-survival (remember the clincher that life in the premoral state of nature is not only nasty but short)—especially in light of Edelman's comments above about how basic evolutionary values ultimately underlie our higher-order moral constructions. We saw above how the intersubjective position can ground itself in the needs, wants, and interests of beings who share biological commonalities of certain physiological functions (hunger, sex, and so on), but surely no value is more integrated into our evolutionary history than survival itself, upon which Hobbes based most of his argument (as did Charles Darwin later, which is no small point).

Hobbes's account of the origins of morality is the prototype of a mythical social contract theory. He imagined that primitive premoral beings, finding themselves in such a state of nature, would create an entire moral and political system to improve the unsavory human condition that is otherwise their fate. This mythical story highlights the observation made in the last chapter of how common sense tends to support the view that we "prescribe" our values even though we "describe" the facts of nature. In the myth, nature is taken as a given while the state (with its moral laws and political sanctions) is an invention to help us deal with it. But just as tables and chairs, once we create them, contribute to our cognitive construction of our experience of them, so too do social institutions, once we create them, contribute to our moral construction of our experience of *them*. We are as much creations of this state as we are creators of it— once we take the broad holistic and historical view that is not so preoccupied with "my own personal" moral experience as it is with moral experience in general.

Just as the philosophy of knowledge has long been preoccupied with the individual's cognitive experience, ethics has long been preoccupied with the individual's moral experience, and the modern version of Hobbes's mythical state of nature is now retold in terms of the game theory that arose out of World War II. This mathematical approach to understanding the relationship between individual psychology and group behavior has had dramatic impact on our political economy, not only in wartime but also in the routine ways we conduct economic life. Mackie offers us an excellent example of how game theory relates to the tradition of Hobbes and Warnock in conceptualizing morality as an institution for counteracting our limited sympathies and ameliorating the human condition.

Consider two fellow soldiers, Smith and Jones, manning positions at the front on top of two hills, the only two members of their squad surviving. Although they cannot see each other, they both know all the following information: They both know that if they both remain and fire at the advancing enemy troops, the enemy will be slowed down sufficiently that reinforcements will arrive in time to give them an 85 percent likelihood of survival. They both also know that if they both decide to run, the oncoming enemy will advance so quickly that their likelihood of survival is only 30 percent. However, each also knows that if he decides to run and the other stays, the one who stays will slow the enemy sufficiently that the one who runs has a 99 percent chance of survival, while the likelihood of survival for the person who remains falls to 10 percent.

What will each man do? Let us analyze the position of each in turn. Smith is unsure what Jones will do, but he knows that in the event that Jones decides to stay, then he (Smith) is better off running, since his chances of survival will increase from 85 to 99 percent. Smith also knows that if Jones decides to run that he (Smith) will again be better off running, since his chances of survival go from 10 to 30 percent. How fortunate for Smith to have an optimal strategy! If Jones stays he should run, and if Jones runs he should run. Whatever Jones does, Smith is better off running, so this is his obvious choice.

Unfortunately, Jones has exactly the same calculation to make. He

The Game Theory Approach to Ethics

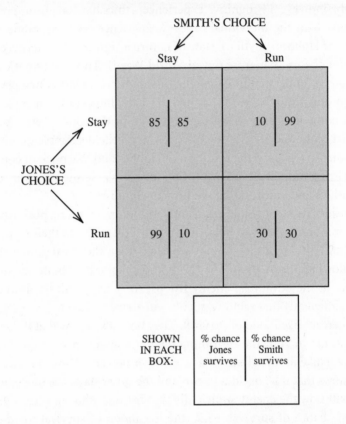

As described in the text, Smith and Jones face a classic "prisoner's dilemma," highlighting the difference between individual and collective rationality. Whether the other runs or stays, each can see that his individual chances of survival are higher if he runs (99% vs. 85% if the other stays, and 30% vs. 10% if the other also runs). But if they act on this individual calculation and both decide to run, their collective chances of survival are less than if they both stay (30% vs. 85%)!

immediately understands that if Smith stays, he (Jones) is better off running, and if Smith runs then he (Jones) is better off running. Following what can only be called a "rational" strategy, both of them run and both of them die.

Assuming evolution has provided brain structures that give both men a preference to live rather than die, what mechanism could be used to get them into the upper left-hand box of the figure instead of the lower right? Hobbes would immediately make a suggestion. He would suggest that both men agree to be chained to their positions on the condition that each knows the other is also chained. By giving up the "freedom" to run, each knows he becomes free to live, thanks to the intervention of whatever Leviathan they devise to organize the chaining.

Of course, physical chains have some impracticalities for the men, and we might ask whether each would be better off having some *psychological* fetters that would keep both of them at their stations. This is presumably precisely the role played by military codes of honor, and the reason why it makes sense during wartime to prefer to be part of a highly disciplined unit rather than a chaotic band of vagabonds. If the 99 percent chance of survival becomes associated with a 100 percent chance of being branded a coward, it may well be that Smith and Jones can get out of their "prisoner's dilemma." (This predicament is known as the prisoner's dilemma because the original scenario had two prisoners, each of whom decides to confess and end up with a longer prison term than they would if neither of them chose to confess, given similar percentages to those shown here.)

One of the most interesting results of the prisoner's dilemma is how clearly it illustrates the difference between what might be called "individual rationality" and what might be called "collective rationality." The individually rational solution sends each man running and dying. We can see why they would prefer to create larger social institutions of military codes of conduct and honor in order to get themselves from the individually rational box (where there is only a 30 percent likelihood of survival) into the collectively rational box (with an 85 percent chance of survival).

Our current view of morality as an institution designed for solving these types of societal problems can afford to be much less harsh than military codes of honor, however. The reason for this can be seen if we imagine that Smith and Jones not only know all of these percentages but also know that if they survive they will play the same "game" over and over. One can imagine that after several repetitions of this game there can be enough presumption of trust that even if both men were no longer bound by military honor and duty they could make a promise to one another to stay and each would trust the other's promise, since they have proven themselves trustworthy so many times before.

In game-theory language, this is called a supergame—a game that is played repeatedly by the players with the same rules each time. To the extent that morality can be understood in these terms, it can only be understood as a "supergame." Although in each individual case, we may have selfish motives and consider ourselves "better off" breaking a promise to a friend, we become bound by the institution of promising in part because we know that there will be many more promises on each side to be shared with our friends, and so we take a less narrow (less individually rational) view of our circumstances. When we see the considerable amount of effort parents make to acculturate their children into the institution of morality (and other related social practices), we might now interpret it as an attempt to engage children through the way they naturally care about how things go for themselves and the world, and to channel future calculations from the individually rational box to the collectively rational box.

But surely the child needs some "natural" tendency to care about the way things turn out to get this whole game off the ground. If someone really did not care what became of themselves or the people around them, it would be hard to imagine how this whole business of morality could ever get started. We can only view the institution of morality as a device for solving the predicament of the human condition if we appreciate that there is some predicament! This was perhaps best captured by H. L. A. Hart when he observed that "men are neither angels nor devils; that makes morality both necessary and possible." If, unlike Hobbes's mythical savages, we did not have any limitations on our sympathies, then morality might not be necessary (presumably, angels do not need such a device). But if we really did not care at *all* about how things turn out—even for our own survival and the survival of those close to us—then morality could never get off the ground.

It is from this observation that we might now better understand two unusual philosophical puzzles that were left open at the beginning of this book. The first came from Hume when he said that "reason is and ought to be the slave of the passions." Far from being a radical *prescription*, we can now understand this statement as an exercise of *description* from Hume, the naturalist. Hume looked around and observed some peculiar human practices: human beings make moral judgments and set up social institutions that embody moral values. When he asked himself, "What is it about humans that makes them do this?" the answer Hume discovered was not that "it is because human beings have the capacity for reason" but rather that "it is because human beings have 'passions'—they *care* about the ways things are for themselves and one another." In this naturalist's voice, Hume's famous dictum is just a reminder that morality only gets off the ground because of certain passions, because of our natural tendency to care about ourselves and those around us. And we now have even better reason to believe this than to believe the logical appeal of Aristotle's claim that "reason by itself moves nothing" (Chapter 2). As we saw with Darwin III, we probably need some passions—some values—to account for the fact that we make *any* type of distinctions, not just moral distinctions.

But this same observation can then give us at least a hint of a solution to the puzzle left from Plato, whose entire cosmology became dominated by a moral "good" with the unusual property that having knowledge of it is automatically to impel desire for it. We can now see why this peculiar property cannot be dismissed as quickly as most moderns believe by relating our Hobbesian game-theoretic myths about the "state of nature" to what we now know about the *real* state of nature. If we return to our biological idiom, we might look back at the nurturing efforts of those parents to build upon their children's natural tendency to care about themselves and about others, and ask where genes and environment fit into this nature-nurture story. Human beings are not the only animals that engage in collective behavior, and evolutionary biologists have demonstrated how some instinct exists not only in apes and chimps but even in birds and bees that might get them at least started toward the solution achieved by Smith and Jones at the front.

Indeed, the animal parallel of Mackie's game-theory example might be the problem faced by a bird as a predator approaches the nest. In this case, its offspring in the nest have a much better chance of surviving if the parent moves forward to distract the enemy, thereby decreasing the chances of its own survival. This is exactly what many species of birds do, in a broken-wing display which lures a hungry predator away from the nest. The parent bird may indeed be caught and killed in this "altruistic" act of pretending to have already been injured, but the offspring, with many of the parent's genes, may survive as a result.

Such altruistic behavior is just what we should expect in an evolutionary drama where the successful continuation of one's genes (even when this contrasts with one's *self*) is a defining rule of the "game." Richard Dawkins popularized this idea—which dates back to Darwin—in his book *The Selfish Gene* (1976). On this view, when members of a wolf pack (who, typically, are all related to one another) behave altruistically toward one another and engage in collective behavior, they do so because of a natural instinct to protect the common genes they share (not to protect a particular wolf such as Smith or Jones). Through this evolutionary concept of kin selection, we may view the natural tendency we have to care about others (the tendency that gets morality off the ground) as *natural* in the real biological sense of the word. Just as a natural instinct to look at faces can be used to harness our plastic brain to an environment filled with crucial information about the emotional expressions of those around us, so can a natural instinct to care about the local cohort (because many of them are our relatives) be used to harness that same plastic brain to care about more remote members of the group (as parents and military drill sergeants each do in their different ways).

We might now reframe H. L. A. Hart's saying about angels and devils in these same biological terms. Although a natural tendency to care for members of our local cohort makes the game of morality possible, an equally natural tendency to make war on more distant cohorts makes morality necessary. Even as we see how wolves within a pack can cooperate and engage in altruistic behavior, we must remember the equally natural outcome when they come upon a competing pack. It is interesting how philosophers have focused so much attention on whether the largest or smallest reference groups are more relevant to ethics (in traditional discussions of objectivity versus subjectivity), when biology provides us with natural value preferences whose intermediate-sized frames of reference hover between the size of the family and the local social group. The view of morality that emerges from Hobbes, Warnock, and game theory alike essentially says that our natural tendency to care locally makes a good (and necessary) starting point, but that without nurturing those tendencies to generalize and broaden them, we will be right back in Hobbes's ugly state of nature.

The arms race has provided a marvelous example. With the two sides not viewing one another sufficiently as part of the same kinship cohort, each side looked at the game-theory chart and saw that whether the other side decided to arm or not arm, "we" would be better off arming! Although a more global view of our human predicament may be emerging with a more species-defined context for who "we" are, it is still an open question as to whether masses of people will die together in the individually rational box, or whether our limited natural tendency to care can be nurtured to include more distant cohorts and move us to the collectively rational box.

Either way, once we have identified such a natural tendency to care about others, it becomes no more mysterious to consider Plato's notion that cognitive experiences can "automatically" impel desires than it is for biologists to consider the sight of water "automatically" to impel desire for it in a thirsty animal, or for the approach of a predator "automatically" to impel a bird to enact a broken-wing display. Plato did not include in his tripartite model of the mind a maternal or paternal instinct to nurture children as part of what he meant by "desire." Whether we would now add this tendency to his faculty of "reason" or "spirit" has, I hope, become irrelevant, as we have seen how virtually all of these distinctions have become blurred during the course of our analysis.

The introduction of the existence of the state and the political economy in a biological discussion of the relative contributions of nature and nurture immediately raises the question as to whether the state is a "natural" entity. This was certainly the view taken in classical Greek thought. Aristotle's observation that "man is by nature a social [sometimes translated as 'political'] animal" is perhaps best understood in context if taken as a piece of biological theory, as the philosopher Jonathan Barnes has suggested. While Aristotle saw Greek city-states as natural entities—as manifestations of human nature—it is difficult to view modern giant countries as "natural" in the way Aristotle did. Certainly research on other primates suggests a smaller cohort than tens or hundreds of millions as the "natural" grouping for beings such as ourselves. But this research perhaps only applies to "man the animal," not "man the thinker," to use that classical distinction.

Our game-theoretical approach has suggested that man the thinker is able to construct social institutions that solve some of the problems of man the animal. These constructions include not just the whole institution of morality but others, including the entire machinery of the state, the political economy with its legislation and market networks, religious institutions, and so forth. Furthermore, while Hobbes's story of the creation of the state was mythical, we know that actual states and economies have real histories that may be studied. That is, although we noted above a dialectic whereby man is both a creator of the state and a creation of it, in this case there is a relatively clear historical order. As that astute observer of political history, Alexis de Tocqueville, noted, before social institutions are causes they are effects. We can study the way these institutions have risen—and fallen—and we can only be impressed by the complexity of levels of interactions. And, as noted at the outset, we should not fail to note how the nature-nurture ethos of the day shapes those institutions that may continue to influence us as creations of those institutions for many years to come.

One lesson we can learn from history is how institutions that survive for millennia must have included plenty of flexible room for nature and nurture, although we may see how different sides of these complex institutions gain more prominence during different periods of history. The Catholic church, for example, has incorporated re-markably diverse—and even antagonistic—cultures and attitudes at various times, and this flexibility helps account for its success. A nature-driven cosmology can emphasize the burden of original sin to power the "device" of morality; a nurture-driven cosmology can emphasize the mind's receptivity in accepting a state of grace. Both sides are integral to any institution that can survive over the long run.

The history of our own political economy is filled with examples of attempts to construct institutions that can incorporate both na-ture and nurture. Classical economists drew the distinction between the two in interesting ways. In his *Principles of Political Economy* (1848), the polymath John Stuart Mill tried to draw a line between those economic laws governing the "production of wealth" (which "partake of the character of physical truths") and those governing the "distribution of wealth" (which are "of human institution solely" and could be made "different, if mankind so chose"). Mill thus might have said that, in a game-theoretic approach, the state of nature which determines the "predicament" includes the "laws of produc-tion" (which are taken by him to be as unalterable as the need for food, shelter, and clothing). The *distribution* of wealth then becomes the sole arena within which economic strategies can be applied.

Mill thus took a rather narrow view of the variables available to us in our game-theoretic attempts to improve the human condition (to increase GNP, in brute economic terms). Indeed, the turbulent history of the world's political economy over the last 150 years might be viewed precisely as a struggle over the definition of what should be taken as a natural given and what should be taken as a variable to work with in the game. After all, it was in part when Adam Smith's free market was combined with the harsh natural laws that Malthus and Ricardo derived from considerations of geometric population growth and linear productivity growth that the seeds of the socialist critique were sown. Marx took a broader view of the variables upon which we may operate than did the classical economists (who only considered wages, profits, rents, and the like). He made the capitalist system itself a variable (and, in his system, a transitory phase in society's long-run evolution).

While the dramatic impact of Marxist ideas was being felt throughout the world, the United States and British economies were influenced more by another proposed set of variables to solve a different picture of the harsh state of nature. This was the picture painted by the influential British economist John Maynard Keynes, who theorized that the laissez-faire free market economy could not guarantee full employment in the long run. His conclusion was that some intervention—through the application of certain "artificial" ideas in government policy—could remedy this failing of the economic state of nature (for example, by attempting to stimulate aggregate demand). Despite almost two decades of anti-Keynesian "supply side" and "monetarist" rhetoric by the heads of state in both countries, the influence of Keynes's view of the available variables can still be seen in our mixed economy—with its combination of laissez-faire "nature" and government-intervention "nurture."

Although these ideas may sound far removed from our earlier discussion, they may be reconnected as soon as we remember how even a discussion of the old question about our knowledge of tables and chairs (Chapter 8) required a reference not only to the thoughts of the educators who put them there but also to the role of the entrepreneur who capitalized on those thoughts. While this type of analysis must result in a great deal of complexity of different levels of interaction, we can at least understand how our knowledge, values, and social institutions are mutually evolving aspects of human living. In our discussion of theories of knowledge we spoke of questioning even our basic analytic definitions (especially during scientific revolutions). In our discussion of moral values and social practices we spoke of questioning even such a widespread value as patriotism. Now, still within our game-theoretic language, we see how even our basic foundational institutions (the geocentric universe, patriotism as a value, capitalism as an economic system) can be analyzed to see if they have become new fetters in a world that is ever-changing in no small part because of these foundational institutions.

Marx saw in class struggle the inevitable process by which the foundations of our political economy could be shaken and improved. Our discussion has pointed to a different kind of struggle to improve our knowledge, our values, and our institutions. This is the struggle to search our successive understandings of the world for those inadequacies and inconsistencies that point to a more complete appreciation of the whole. It is a struggle to be carried out by our powers of reason. Let us take a closer look at how this process works.

18

Natural Law, Nurtural Law

While David Hume recognized that some natural sympathy for others is required to get the whole business of morality going, his descriptive, naturalist approach to ethics left him with an unusual problem. Among the ethical ways of behaving, there are some toward which people are certainly naturally inclined, such as kindness to children. But, as he observed the various ways in which people engage in moral behavior, Hume also saw people identifying ethical standards which humans are most certainly not inclined toward, his most famous example being principles of justice. Hume argues persuasively that there is nothing particularly "natural" about our notions of justice. Although he turned Plato's theory on its head and made reason the slave of the passions, Hume understood that virtues such as truthfulness and courage are generally not naturally attractive: that people need to *think* (use their faculty of reason) to see why these are good things. It therefore became important to Hume to distinguish between "natural sentiments" and "learned sentiments" that together account for moral behavior. He thus distinguished "natural virtues" (such as compassion for children) from "artificial virtues" (such as a sense of justice), the latter of which—although not straight-off attractive—can be learned as being good for society and then inculcated ("nurtured") into individuals.

In modern times, we might want to be more precise and separate the particular moral distinctions people make (which Hume found to be largely artificial) from whatever instinct people might have to be moral in the first place. Such an instinct would by hypothesis be "natural," but it would not guide particular behaviors in one way as opposed to another. Just as we possess an instinct to acquire some language—but whether this will be Chinese or English depends on our upbringing—so we could still allow for, say, an instinct to want to be the most virtuous member of the tribe, with our upbringing here determining whether this means being the most skillful warrior or the most giving philanthropist. Hume's observations still hold true for the particular distinctions people adopt about justice, even if it may be only natural to want to compete to be the most (or least) just individual around.

If, in keeping with this descriptive, naturalistic approach to the problem of morality, we were to survey both the scholarly literature and the popular press to catalogue what sorts of moral distinctions people make, we would not be surprised to find that most moral virtues are artificial in Hume's sense. Indeed, this is probably why we make such a point of saying that truthfulness, courage, and a sense of justice are "virtuous." If we were readily inclined to act that way anyway, we would not need to give such values so much "moral support"!

There is, in fact, an entire linguistic approach to understanding morality that looks at the ways in which we use words like "good" and "bad" simply as a reinforcement mechanism to show our approval or disapproval of various behaviors. Indeed, at the extreme, the language used in religious writings can, as the contemporary moral philosopher Alasdair MacIntyre has shown, be understood as an attempt to engage the powerful "objective" force that only a God could give moral precepts. In this strategy of shaping individual sentiments and nurturing our passions in an unnatural direction (perhaps just short of the all-powerful Leviathan imposing physical chains), an all-powerful God can be engaged to apply the strongest of psychological fetters to help move us from the individually rational to the collectively rational box in our game-theory model (which we have taken to be the point of morality).

This view of the moral force of religious language can operate in a variety of ways. A wrathful, vengeful God such as is sometimes portrayed in the Old Testament can, for example, secure a deep sense of the artificial virtue of justice in any believer raised to fear that all-powerful God. Another strategy can be found in the idea of Christian love. Everything we know about human (and animal) nature tells us that loving one's enemies is an unnatural or artificial virtue. However, as part of a religious cosmology, children can be trained to appreciate a connection between love for one's enemies and an eternal state of grace. Yet another contrasting religious strategy may be found in Tibetan Buddhism. The Dalai Lama maintains that the natural compassion that we have for those closest to us (which gets morality off the ground) can be generalized to all beings in the world through specific meditative practices. That is, we can, through those ancient practices, shape our plastic minds to take advantage of the seed we have of a natural virtue, nurture that seed (artificially), and thereby come to operate in the collectively rational box.

The picture that begins to emerge here is of a great *diversity* of strategies that have evolved in society to nurture children in those artificial virtues that are needed to counteract the social problems created by nature. We tell small children stories about how the tortoise beats the hare in the race, not because we want them later in life to put their money on the tortoise! It just does not come naturally to small children to be slow and steady, and experience has shown that slow and steady leads to more productive outcomes. The story of the tortoise and the hare is a parable—not unlike parables in many religions—a prototype of the artifice that defines the entire institution of morality. Through sometimes subtle application of this artifice we introduce into society and, through nurture into our children, inclinations for humans to act in such ways that will make things go better for all of us ("better," again, in the naturalistic, nonmoral sense of less frustration of human needs and desires, or higher GNP in economic terms).

But what distinguishes morality from other forms of indoctrination, be it indoctrination into good character and personality development or indoctrination into various forms of prejudice and bigotry? We also nurture children to enjoy certain types of art and music which might not otherwise come naturally. How does ethics differ from aesthetics?

At this point, it should not be surprising that the answer here will de-emphasize old-fashioned distinctions between "character development" and "moral development" and "aesthetic development." Interestingly, when we look back to the classical literature, these distinctions did not exist. Indeed, there is some question as to whether you could even express these distinctions in Greek or Latin, so that Aristotle saw himself as addressing the unitary question, "How am I to make a satisfactory job of my life?" and Cicero could use the same language to instruct children not to cheat in their lessons and not to trail their togas on the ground.

How can we now account for the importance attached to these distinctions? The modern version of these distinctions may again be attributed to Kant, who saw in different elements of our faculty of reason different applications to cognitive, moral, and aesthetic judgments. Kant's ethics is famous for maintaining, in contrast to Hume's, that the making of moral distinctions is wholly explicable because people are rational beings. In sharp contrast also to classical thought, Kant maintained that there is a huge gulf between moral reasons and other reasons, and his famous categorical imperative made "universalizability" a defining feature of moral maxims ever since.

Certainly in our common usage, part of what we mean by ethics is rules of behavior that, if they apply to one person, apply equally to anyone else in the same situation. We see here another hint of Kant's emphasis on the distinction between form and content, since he believed reason could identify the form of a moral maxim by checking that it can consistently be universalized to anyone in the same situation. Like his focus on the way the mind provided the form of any possible experience (through the conceptual categories of the faculty of understanding), Kant believed that our mind determines the form of any morality. Indeed, his vagueness on the content side of the equation (paralleling his vagueness in the cognitive realm about exactly how the faculty of sensibility "receives" its sensory impressions from the world-as-it-really-is) has been blamed for much of the moral lunacy in the world ever since Adolf Eichmann responded with the answer "Kant's categorical imperative" to the frustrated prosecutor's question about the moral theory to which he subscribed.

Although universalizability is one of the most important criteria we use to identify moral claims (in contrast to aesthetic and other claims), we have already seen how blurred is the distinction between form and content. Kant did not say anything about how we should pick the maxims to which we would apply his test of universalizability. Although he seems to have thought that the maxim for any action I might consider is obvious, disagreements are often precisely about what the relevant maxim is. This is especially important if we take most of these moral judgments to be artificial devices that we have created to try to nurture young minds to improve society. A wide diversity of notions about "justice" have been propagated in different societies throughout the world. How can we determine the proper content of morality once we understand the entire institution as an artificial device for ameliorating the human condition?

If we look back at Chapter 16, we remember how something that started to look like a theory of morality based on natural law (with values determined by the biological givens of people's needs, wants, and interests) actually revealed itself to be much more integrative of nature and nurture. This "nurtural-law theory" no longer takes values as self-evident, as Thomas Jefferson did, but seeks new and better values, many of which are quite *un*natural. We made the point that, just as science is the search for things worth naming, ethics is the search for things worth valuing, and so we need some criterion for what makes better values better.

In the choice between an *external* solution to this problem (which would look outside of the practice of morality for some yardstick) and an *internal* solution (which remains within the institution of morality), we can only opt for the latter approach. Although the internal approach raises fears of circularity, the external approach is condemned to failure by the absence of any place outside the institution appropriate to search. We have already seen how such an internal solution can be generated. We must apply our faculty of reason to search our moral precepts for standards by which we intend to judge even those precepts themselves. In order to make this (counterintuitive) process work, we need to integrate what we have learned about our construction of knowledge with our new understanding of the structure of morality. In so doing, all of the academic distinctions (between ethics, aesthetics, the philosophy of knowledge, and so on) will break down, as I shall demonstrate using one simple example in the next chapter.

But first let us consider how our powers of reason (which are meant to fuel successive and improving conceptions of our knowledge and our values) relate to the "natural law/nurtural law" ideas that have been presented here. If traditional natural law theories of the sort that fueled the ideas of the framers of the U. S. Declaration of Independence and Constitution were still in vogue today, there can be no question that they would have to be based—however indirectly—on the principles of evolution. Today we say that evolution is what "gave rise to our nature." If reason is to be applied to nurture our ideas, moral values, and social institutions, what can we say about the relationship between reason and evolution?

I shall make two points about this relationship that have direct bearing on how reason can participate in that noble struggle, described at the end of the preceding chapter, to improve our knowledge, our values, and our social institutions. The first concerns the sense in which we would or would not want to consider reason itself a "natural" product of evolution. The second concerns the actual historical appearance of reason in evolution.

One of the most counterintuitive things about our human capacity to make rational inferences and deductions is how *irrelevant* this has been to evolution. Faced with some practical problem to solve, humans as products of evolution will generally apply some reasoning process that balances such factors as the speed with which a decision must be made and the dangerousness of the situation with the metabolic (energy) costs of this decision-making process. This process has very little to do with logical syllogisms like "if A is B and B is C, then A is C" and other similar products of our faculty of reason.

Let us take an example. Consider for a moment the two thieves breaking into the house who, when captured, ended up in the original prisoner's dilemma. Both confessed and ended up getting a longer prison term because each "rationally" calculated that whatever the other does, he would get less jail time by confessing. We saw how this individually rational outcome can only be prevented through some device that can move both self-interested men into the collectively rational box.

But surely they would be better off not getting caught in the first place! Imagine that what happened instead was that each of the men saw someone coming down the street as they were breaking into the house. They cannot really see the person closely, but they see he is wearing blue. The insightful Harvard psychologist Brendan Maher has given a wonderful description of how evolution has shaped the thought processes that would go through their minds, assuming that getting captured is a disadvantageous outcome. Would they reason: policemen wear blue—this man is wearing blue—but that does not necessarily mean he is a policeman—but perhaps we could calculate the rational odds based on some estimate of how often non-policemen wear blue—but . . . ?

Assuming this kind of reasoning costs both metabolic energy and time, the kind of adaptive behavior selected by evolution would be very different if getting captured would be a fatal outcome. As Maher shows in several such examples, the *adaptive* syllogism in such a case would be: policemen wear blue—this man is wearing blue—he must be a policeman—run for it!

But if rationality in the strict traditional sense is not really relevant to evolution, then how did it happen to evolve? The answer is that this analysis shows only how our natural *individual* rationality is irrelevant to adaptive behavior. As soon as the burglars have been caught and put into the prisoner's dilemma, they will need to apply their faculty of reason to get them into the unnatural but more adaptive *collectively* rational box. Put another way, we might say that in Hobbes's gruesome mythical state of nature (which for our purposes presumes that evolution of the species somehow impossibly occurred in isolation from the evolution of social institutions such as morality and the state), the faculty of reason in these unfortunate beasts is in *its* "natural state." Until the further appearance of some artificial kinds of reasoning (including, perhaps, not only institutions like promise-keeping and justice but even the syllogisms of logic!), life in the individually rational box will continue to be "solitary, poor, nasty, brutish, and short"!

Although Hobbes's state of nature is only a myth, our evolutionary variation of it immediately raises our next question concerning the actual historical appearance of such an artificial thing as the capacity for collective rationality, logical syllogisms, and so forth. What is important here is not so much a precise date (was it 50,000 or 200 million years ago?), since we can only assume that such a capacity evolved gradually once primitive consciousness arose and conscious beings had time to evolve language and construct conceptual maps of their conceptual maps of their . . . (Edelman, by the way, estimates that primitive consciousness arose about 300 million years ago.)

What is very important is the idea that the appearance of such a capacity would mark a qualitative shift in the strategies that might be employed to play the game of nature. Many thinkers have offered frameworks for how to conceptualize this qualitative shift (E. O. Wilson's theory of sociobiology is a powerful example), but I have chosen the framework offered by the developer of the polio vaccine, Dr. Jonas Salk, for reasons that will become obvious in the final chapter. Salk's language initially divides the history of evolution to date into three periods: prebiological, biological, and metabiological. *Prebiological evolution* is the evolution of matter itself, from the big bang to the evolution of the stars, planets, heavy elements, and so on. Prebiological evolution is extremely slow.

There is no precise point at which *biological evolution* appeared and became superimposed on the prebiological evolution of matter (which continues along in the "background" against its much faster-paced new counterpart). Some thinkers have placed the start of biological evolution at the point when matter became complex enough to give rise to self-replicating systems; others have focused more on the emergence of the transfer of information between systems. It is in any case clear—not just in our wheel of knowledge about them

Salk's View of the Evolution of Evolution

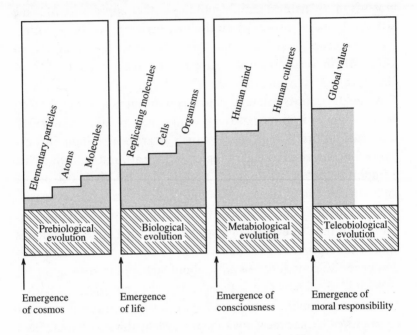

but in historical actuality—that biology ultimately emerged from physics and chemistry, and this more rapid biological evolution eventually gave rise to *Homo sapiens.*

Just as there is no sharp dividing point when life emerged to create a new biological evolution from out of prebiological evolution, so there is no sharp dividing point when the human mind emerged to create a new *metabiological evolution.* Metabiological evolution is Salk's term for the even more rapid and more complex stage in which human thoughts interact with (and influence) the biological world. (Remember that ideas about logic emerge from biology in our wheel and do not just support mathematics at the base of some reductionistic "tower"!) When *Homo sapiens* began digging irrigation ditches to enable otherwise fragile species of crops to flourish and provide food, they were participating in a new process through which things interact with *thoughts* (and some things "get there" because people think they "should be there"). As before, the earlier stage (in this case, of biological evolution) also continues while the newer, faster-paced stage is superimposed upon it.

Salk notes that this progression is itself a natural evolutionary process—a process by which evolution has undergone evolution. Indeed, the movement from prebiological to biological to metabiological evolution is perhaps the ultimate example of that dialectical movement of the whole that Hegel tried to capture with his metaphor of bud-to-blossom-to-fruit. But, Salk suggests, as part of a continuing evolutionary process itself, evolution may continue to evolve beyond the metabiological stage. This idea relates directly to our historical variation of Hobbes's myth. It is one thing for human thought to emerge and begin to mix itself with the realities of biological life. It is quite another to postulate that these thoughts can begin to *improve* upon nature, to create artificial devices that employ collectively rational processes to solve some of the new "prisoner's dilemma" problems that—like the arms race and global pollution—threaten our very survival.

Salk has proposed the name *teleobiological evolution* for this new stage which, as the name implies (*telos* means "goal" or "end"), would include some *directionality*, some responsibility for shaping the course of metabiological evolution. It is sobering to think that the increased pace of this stage might again make its predecessor look almost stationary by comparison. He suggests that if our survival depends upon our success in taking this responsibility, then what hangs in the balance is the *success of evolution itself*—whether evolution can evolve to solve the problems created by the greatly accelerated rate of metabiological evolution.

This challenge to create a nurtural-law theory of values in response to the predicaments left us by natural law will be taken up in the final chapter. But first I promised a simple example of how this analysis has broken down so many of the distinctions once thought important to whatever reason we may apply to these problems. Let us return briefly to the "basics."

19

Lessons from an Optical Illusion

In discussing the origins of modern Western philosophy in Chapter 3, we saw how important was Descartes's distinction between thoughts and things to all of the thinkers who have followed. Although introspection makes such a distinction fairly obvious, the existence of the distinction is especially apparent in our experiences with optical illusions, where by hypothesis our *experience* of the thing has characteristics different from the *thing* itself. With the help of the Müller-Lyer arrows, we explored Kant's emphasis on the interaction between ourselves and the objects we experience in the process of coming to have knowledge of those objects. By keeping our focus on this interaction between ourselves and the objects, it became easy to move from the philosophical idiom of Part One to the biological idiom of Part Two, where this interaction became understood in terms of the actual neural mechanisms through which our brain and the world itself are mutually shaped in the course of coming to have experience of "things."

With this biological view in mind, we might now ask what it is about our visual input system that causes us to misinterpret the world in cases such as the Müller-Lyer illusion. Although physiologists have tried to explain this particular illusion as being caused either by some optical effect in the eye or by a disturbance of the signals from the retina created by the fins on the arrows, the best

R. L. Gregory's Explanation of the Müller-Lyer Illusion

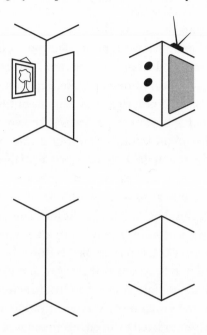

R. L. Gregory has proposed that this illusion is caused through the combination of two (usually correct) assumptions made by the visual input system in analyzing such patterns. First, the depth perspective cues shown in the figure lead the visual input system to the conclusion that the left arrow shaft is behind its fins while the right arrow shaft is in front of its fins. Second, the system also makes a size-constancy assumption: things do not actually get smaller as they move farther away (they just look smaller). When these two assumptions are taken together, the brain's visual input system concludes that, because the left arrow shaft is farther away, it must be bigger than the right one, since both cast the same size image on the retina.

explanation of this particular illusion was put forth in the 1960s by Richard L. Gregory, who actually became one of the pioneers of artificial intelligence in part by wondering whether robots could learn to handle the way we do certain aspects of perceptions such as this. As demonstrated in the figure on this page, Gregory realized that the fins on the shafts of the Müller-Lyer arrows introduce information that the visual system uses in processing depth perspective and size-constancy scaling. A brief explanation of each of these two features—depth perspective and size-constancy scaling—will help us understand how this illusion works.

In an environment filled with rectangular rooms, buildings, furniture, boxes, TV sets, and so on (the "natural" environment of most of us), right angles appear obtuse or acute, depending upon whether the corner in question is pointing away from us or toward us, respectively. This element of linear perspective was known to ancient Greek and Roman artists in an intuitive way, but it was first "discovered" (intellectually) in the Italian Renaissance by Filippo Brunellesco (1377–1446), who was able to explain mathematically the curious distortions in so many classical works, such as the unrealistic-appearing architectural background in Rome's famous Arch of Titus. Children raised in a rectilinear environment like ours spend their early years exploring these rectangular features of the world with their mouths and hands as well as their eyes, and their plastic visual input systems shape themselves to these realities using such cues to analyze depth in the visual field. Very early in life, our visual input systems have accommodated to an environment where obtuse angles are analyzed as 90° corners pointing away from us and acute angles as 90° corners pointing toward us.

In exploring the artistic portrayal of three-dimensional space, Brunellesco also codified another quality of linear perspective: the degree to which objects of the same size seem smaller the farther they are from the observer. Since "things" do not actually get smaller as they move away from us, our brains also accommodate to the size constancy of objects regardless of the actual size of the retinal image: a six-foot-tall man does not appear to shrink by 50 percent as he walks from ten to twenty feet away. Similarly, you see your two hands as equal in size when holding one twice as far from your face as the other, so long as they are held a few inches apart. As you move one behind the other, the half-sized retinal image of the farther hand reveals your brain's application of this size-constancy scaling (try it!).

So, in the case of the illusion, our visual system "sees" two lines casting the same size image on our retinas, one of which is presumed to be closer than the other (because of the depth cues from the fins).

Since they cast the same size image on our retinas, the closer one must be smaller and the farther one bigger; this is the information processed by our visual input system and presented to our brain's central processors. We therefore see the arrow shaft with obtuse fins as longer and the one with acute fins as shorter. This processing system usually employs such subtle cues to construct a very "accurate" picture of our rectangular world, but we only discover this whole intricate process when it gets fooled by the illusion.

It is tempting to call this phenomenon of the accommodation of our plastic visual input analyzers to the realities of depth perspective and size constancy in our world a "natural phenomenon"—except for the simple fact that such rectilinear lines do not exist *in nature!* One may therefore ask what becomes of the visual input system in the brains of children who are raised in caves, or in round huts with no rectangular cues for depth perspective? Studies have in fact been done on a number of such "circular cultures," where huts are round, land is plowed in curves rather than straight furrows, children's gifts are not wrapped in boxes, and few possessions have corners or straight lines. These people rarely experience the Müller-Lyer arrows as an illusion, and when they do it is much less of a distortion. In fact, in the 1970s the anthropologist J. B. Deregowsky, who studied one of these cultures, the Zulus, showed that those Zulus who do not experience the Müller-Lyer illusion also show little or no perceived depth in the figures even when the figures are presented under special conditions that give them added depth perspective for Westerners. A number of interesting results may be drawn from these observations.

For one thing, this anthropological discovery certainly argues against the physiologist's nature-side explanation that the illusion is caused by some genetically determined feature of the eye or retina. It in fact lends strong support for Gregory's nurture-side explanation of this illusion (whose fame has so outlasted that of the German psychologist Franz Carl Müller-Lyer [1857–1916] who discovered it). The illusion highlights what we saw in Chapter 11 about the evolutionary need of perceptual systems to wire themselves up quickly during brief critical periods early in life so that maximum speed and efficiency can be achieved. It also reminds us that sensory data are not "raw" but are processed by our faculty of sensibility using some simple assumptions (like size constancy and depth perspective cues) that can lead to errors. Our experience with this illusion exposes this elaborate processing even of sensory data and forever puts an end to the *tabula rasa* notion through which Locke and Hume wanted to guarantee the certainty of the senses as a premise for guaranteeing the certainty of our conceptual beliefs.

But more than that, we now have a very concrete example of what in Chapter 8 we called the "third direction" of the relationship between thoughts and things. Certainly, in keeping with Kant's original observations, the Müller-Lyer arrows on the page contribute quite a bit of basic sensory information about themselves to our thoughts about them. In the other direction, we are also reminded how the structure of our thought (with its sophisticated input analyzing network in the visual cortex) contributes to our experience of those things—in this case, giving rise to the optical illusion. But what we see here in the clearest possible terms is also the contribution of the third direction: the contribution of thoughts (not necessarily our own) in *giving rise to* those rectangular rooms, buildings, gift boxes, and pieces of furniture that in turn gave rise to the shape of those neural structures in the first place. It is, after all, only because people had *ideas*—ideas about architecture and the best way to build a table or a chair—that our Western rectangular world contains useful information about depth perception in the form of rectilinear lines that may appear obtuse *or* acute.

But what kind of judgments are these that led to this rectilinear culture that led to our neural structures that led to our perception of this illusion? The answer is that they were *aesthetic* judgments. It is, after all, a matter of aesthetics how we build our buildings and our furniture and what sorts of packaging we use when giving presents to children. Thus, although many writers have commented on the blurred boundary between aesthetics and ethics, one lesson we can draw from the Müller-Lyer illusion is that the boundary between aesthetics and the philosophy of knowledge is likewise blurred. As we have just seen, our aesthetic judgments can shape our neural cognitive apparatus to affect dramatically even our perceptual experiences of the world.

This blurred boundary between aesthetics (where beauty is so often taken to be "subjective") and the philosophy of knowledge (where facts are so often taken as "objective") should almost be expected at this point in our discussion. It is a natural consequence of the intersubjective position where, as we saw in both the cognitive and moral realms, two different "becauses" may both hold simultaneously. We have (at least some of) the thoughts we do *because* we live in a rectilinear culture, *and* we live in a rectilinear culture *because* of thoughts that people have. This dialectic is embedded in a holistic view of the complex relationships between "thoughts" and "things," where questions about our knowledge can only be answered with reference to people present and past (with their cognitive, moral, and aesthetic sensibilities) who brought those rectilinear buildings into our world.

Indeed, it is not just the shape of the buildings in our culture that participates in this dialectical relationship with the structure of our thoughts. Some of the rectilinear buildings in Western culture function as courts of justice. In the opening of the last chapter, we saw how Hume maintained that justice is an "artificial virtue." We might now say that Hume was simply drawing our attention to the similarity between the "unnatural" architectural form of these buildings and the "unnatural" values they are built to preserve. To the extent that both are part of our history, however, both participate in this dialectic and shape our aesthetic and moral judgments even as we apply those judgments in experiencing (and often criticizing) those aesthetic and moral features that we "find" in the world.

Recall what J. L. Austin once said of over-simplification: "One might be tempted to call [it] the occupational disease of philosophers if it were not their occupation." One way to understand how naturally so many distinctions in contemporary philosophy seem to have eroded is to appreciate our at once oversimplified and highly complex assumption that there is a single coherent world which participates in each of our individual minds and which contains great buildings that are *simultaneously* cognitive (with their temporality, spatiality, and object-ness), moral (with their application of highly elaborated rules of justice), and aesthetic (with their rectilinear architecture). Although a great deal has been gained by intellectually carving up these single entities into separate disciplines for study (philosophy of knowledge, ethics, aesthetics), the time has come to appreciate how much has also been lost as the sum of these parts has become less than the whole.

It was easier for Hegel to write about the moral sphere, since his use of the German word *Sittlichkeit* was able to express the complex of customs, rituals, rules, and social practices which constitute the substance of society and so constitute a part of each of us. *Sittlichkeit*, for Hegel, was a "natural" synthesis of our internal moral sense and our external social world, and so harked back to the classical Greek conception of a "way of life" that did not emphasize these more modern distinctions.

Here we can see more explicitly how the identical dialectic is contained in the relationship between "thoughts in our head and things in the world" and the relationship between "values in our head and social practices in the world." All three directions apply to both! The contemporary philosopher John Rawls has discussed the latter relationship between our moral sense and our social world in terms of what he calls a "reflective equilibrium" of principles and practices. An example from my observations of the training of psychiatrists will demonstrate how this equilibrium gets established through the interplay of all three "directions."

By watching the moral development of young people who plan to become psychiatrists, it becomes easy to see the relationship between values and social practices in action, whether we discuss this relationship in terms of Hegel's *Sittlichkeit*, Rawls's reflective equilibrium, or our "three directions." A difficult moral issue in psychiatry concerns suicide, for example. The way psychiatrists-in-training experience the suicidal behavior of acutely depressed patients will of course have been shaped by the social practices of their respective cultures and upbringings. This is the first direction: a "grid" of values shaped early in life will be "thrown over" any experience with suicidal patients and contribute to the way such episodes are experienced by a student doctor.

But repeated experiences with actual depressed patients who attempt or commit suicide inevitably reshape the moral grids these budding psychiatrists apply when confronting these difficult situations. Often, a young person who enters the field believing in an absolute prohibition of suicide will soften this moral stance after beginning to appreciate the experiences of patients whose intractable suffering really can begin to seem like a fate worse than death. Similarly, a student who enters the field with a permissive view of suicide as a rational option for all people will often pull back from this casual stance after experiencing the tragic suicide of patients whose depression could probably have responded fully to treatment, with the return to a fulfilling life and loving family.

With these real-world encounters contributing to a changed moral experience of suicide, these young people often—upon reflection—change their own articulation of their earlier, more primitive moral principles ("suicide is never permissible" or "suicide is always permissible"). This is the second direction wherein social practices contribute to the shape of our values, and it highlights Rawls's choice of the term "reflective equilibrium." The equilibrium gets re-established through reflection, and it is ever-evolving: their newly refined moral principles likewise change the way these doctors will experience the next encounter with a suicidal patient.

But the *Sittlichkeit* of the morality of suicide also includes certain social rules that are actually codified in the ethical standards of the psychiatric profession and in the laws that regulate the medical profession. Like the tenets of established religions, written principles of ethics and laws regulating medicine are part of the social practices that shape a culture's view of suicide, and they are also ever-evolving. Psychiatrists who have had powerful personal experiences will lobby to change the ethical code of the profession and laws that relate to these issues, and years later these revised standards will be part of the social practices that shape the moral attitudes of mentors and trainees alike. We again remember Tocqueville's summary of this third direction: that social institutions are effects before they are causes. *People*—their ideas and their actions—bring these elements of our culture into existence.

The same dynamic has presumably been at work in religious institutions, where we can imagine it was high priests' or rabbis' direct experiences with suicidal members of the flock that caused them to find all of those "loopholes" in the absolute prohibition of suicide. This third direction then again feeds back, as the softening of the tenet that suicide unequivocally prevents entry to a next life becomes part of a less-absolutist religious culture in which the values of young people will be shaped. The ever-evolving nature of these relationships will play a central role when we return, in conclusion, to our question of how to evaluate proposed artificial values.

But it is worth first noting how all these dynamic relationships between thoughts and things *and* between moral values and social practices shed fresh light on our earlier discussion of objectivity and relativism and the distinction between facts and values. We have already seen how the modern mind is predisposed to the philosophically incoherent combination of cognitive objectivity and moral relativism. In the language of Chapter 16, we might say that as we consider the multiple overlapping spheres to which our claims about knowledge and values are "relative," we are predisposed to think that most facts have a very large reference group and most values have comparatively small reference groups. Of course, anthropologists remind us that there are times when smaller groups constitute the frame of reference for "facts," since anthropologists tend to focus on aspects of experience where such differences are seen between cultures. But most people feel more comfortable with the notion that a "fact," to count as a "fact," must be binding on all human beings.

We can begin to see how the dialectic between moral values and social practices is not merely parallel but *identical* to the dialectic between thoughts and things when we remember that at least some values only count as (intersubjective) values because they are also binding on all human beings. As we have seen, Kant's universalizability criterion made this true of all moral values, but we need not be as strict as Kant in this sense. We may identify existing values or even invent new values with reference groups of various sizes, as when we might try to nurture new values to protect the minority group interests in all cultures or help the two sexes better appreciate the unique needs of the other. In fact, what we might picture is some kind of sliding scale of reference group sizes, with the smallest at one end (bow three times on passing the chief's son) and the largest at the other (try to keep your promises). This sliding scale idea is equally applicable to our theory of knowledge. We can imagine examples of "knowledge" with the smallest frame of reference at one end (Everest is the Earth-Mother Goddess) and the largest at the other (Everest is the highest mountain on Earth).

If we focus on the smallest-reference-group end of these scales, we can immediately raise many interesting questions about various unusual or even "pathological" experiences, such as eccentric aesthetic tastes, unusual religious cults, or fixed psychotic delusions. If we focus on the largest-reference-group end of these scales, we can immediately raise many interesting questions about the true nature of intersubjectivity in both the cognitive and moral spheres. But in this larger context, we must remember that even at the very end of the largest-reference-group side of these scales, we are not talking about "objective truths." Indeed, in suggesting that there exist both cognitive and moral intersubjective truths, I am emphasizing a parallel between the facts and values at this end of the scale *not* because there are some values that are objective in the way most people take all facts to be, but because even the most indubitable facts (even a priori, analytic facts) are a little bit like the way many people take all values to be. They are "relative to us," and not "necessary" truths in any meaningful sense. I am told that the Indian subcontinent is still slowly colliding with Asia so that some of Everest's Himalayan competitors may someday outstrip it for the "factual" title of Earth's highest mountain.

With this sliding scale idea in mind, we might reframe our modern prejudices to say that, even after the largest-reference-group end of the scales has been recognized as "intersubjectivity" rather than "objectivity," there is still what might be called a *quantitative* (if no longer qualitative) difference between facts and values. That is, the majority of facts are taken to fall on the intersubjective side of the cognitive scale (with only some anthropologists and psychologists trying to remind us of the surprising amount of cognitive life that operates with smaller frames of reference), while the majority of values are taken to fall on the opposite side of the moral scale (where truly intersubjective values are few in number and also general in application, leaving much room for different moral ways of living in the world).

What we must not forget, however, is one last lesson from our Müller-Lyer illusion. There may come a day when rectilinear cultures have so influenced the last remaining circular cultures that every living human sees the Müller-Lyer arrows as unequal. In such a world, we can only imagine that the *universal* processing mechanisms of the visual input analyzers of every human being would be taken as very strong evidence that this is the "natural" way for human brains to process these types of shapes.

The lesson we must draw is that even a standard such as this— that *every* human brain works that way—cannot be taken as a given of "nature." In the modern world, not everything that is universal is natural. If you ever begin to doubt this point, made much more eloquently by Hilary Putnam, just think for a minute about Coca-Cola! Thus, if a day comes that one view of justice has so dominated the world that all cultures adopt it and nurture their young in it, neither could we take this as evidence for that being a "natural" value. Neither facts nor values are matters of majority rule! This is why we need to use our faculty of reason to investigate our facts and values so carefully for those inconsistencies and inadequacies that can help us *improve* them both. When it comes to the Müller-Lyer illusion, we measure the lines and quickly understand that our perception of the lines is misleading. This misperception then no longer plays a role in our knowledge: we *know* them to be equal even when we perceive them to be unequal. And, having noticed this inconsistency between our measurements and our perceptions, we are on guard the next time we see a similar shape and we bring a ruler to measure the lines *before* our misperception can get us into any trouble.

Let us conclude by considering what sort of ruler we could use when measuring our conceptions of justice rather than our perceptions of the Müller-Lyer arrows.

20

Nurture Improving upon Nature

A philosopher friend of mine, Ken Westphal, once told me that if I really want to understand the relationships between the natural world and our moral and aesthetic judgments about it, then I should go to a county fair at which they are judging the competition for the "best" sheepdog. The sheepdogs on display are unquestionably part of nature. They are real animals that have grown out of the real world. However, as you listen to the discussion of whether each is a "good" specimen of a sheepdog, you soon learn that many aspects of a "good" sheepdog, such as the shiny color of its coat or the shape of its snout, are not natural at all but are the outcome of years of artificial breeding—specifically because of the ideas people have chosen to hold about what constitutes a "good" sheepdog.

The breeding of animals according to particular ideas we humans choose to hold about our fellow creatures is part of what Darwin called artificial selection—in contrast with the process of natural selection that carried on for millions of years until we humans began to mix our artificial ideas about nature with nature itself (what Salk called the emergence of metabiological evolution). When the standard for whether our artificial ideas about what constitutes a better sheepdog is the judging at a county fair, we just apply the same "unnatural" criteria for the best result that we used to breed the dogs in the first place. We mix our ideas about good sheepdogs with the nature of these animals only to then judge them against these same ideas.

But how could we ever decide whether the prize-winning dogs are really *better* in any meaningful sense? This question cuts to the heart of whether it makes any sense even to talk about ethics as a search for better values, since we need some (value) standard by which to judge a better value, and so we confront the presumed circularity of our internal foundational approach in our moral philosophy just as we did in our philosophy of knowledge.

Just as our search for better facts within the context of our knowledge (itself placed within the broader context of *living*) turned out not to be circular after all, so can our search for better values be understood by again taking the broadest possible perspective. In both cases, the process involves the application of our faculty of reason and a certain amount of trial and error. In the case of the sheepdogs, for example, we could *stop* "nurturing" those characteristics we chose to value, return our prize breeds to nature, and see what happens. If those artificially nurtured shiny coats really turn out to be better in some natural (adaptive) sense, we would not see a duller "wild-type" color return in a very small number of generations (as would almost certainly be the case in this example). Even without doing this experiment, we might suspect that our ideas about pure-bred sheepdogs do not produce "better" animals in any sense beyond winning blue ribbons at county fairs. The notorious hip malformations of most pure-breeds would probably give us a strong hint that we have not improved upon nature's version by manipulating breeding patterns with an eye toward a shinier coat.

Although we could actually do this sheepdog experiment, we can only imagine whether other artificial ideas, such as justice or promise-keeping, would quickly disappear if they were not constantly and continually nurtured—as is typically assumed in fictional accounts of societies run by asocialized and usually rather ruthless children. Interestingly, these brutal, fictionalized accounts usually try to imagine a world where values are *not* artificial, but where people carry on in the same "thoughtless" way that nature does. This point was perhaps best made by the nineteenth-century

German philosopher Friedrich Nietzsche, who is often credited with the first attack on many of the distinctions that have dissolved here (objectivity versus subjectivity, fact versus value, and so on). So, in *Beyond Good and Evil* (1886), Nietzsche writes: "You want to *live* 'according to nature'? O you noble Stoics, what fraudulent words! Think of a being such as nature is, prodigal beyond measure, indifferent beyond measure, without mercy or justice . . . how *could* you live according to such indifference?" [emphasis in original].

Shortly after this passage, Nietzsche goes on to add that if to "live according to nature" we mean to include our human ideas as part of nature, then, he asks, "How could you *not* do that? Why make a principle of what you yourselves are and must be?" This is ultimately the paradox that forces us away from old-style natural law theories to a nurtural law theory in which our artificial values can be mixed not only with nature (with children in particular) but also with the social institutions that so influence our lives, and so can improve (or worsen) our human predicament.

There are many metaphors for this process by which humans have started with something "natural" and then mixed their "unnatural" ideas with it to try to shape nature for their own purposes. From the most atrocious applications of eugenics by fascist regimes during World War II to the most arduous attempts to achieve extremes of human compassion in Tibetan Buddhism (where elaborate mental exercises are performed to try to enlarge our "natural seed" of compassion), human beings have found no shortage of ways to mix the ideas that have arisen from *their* nature in the process of nurturing nature itself. Although the clearest modern examples of this process range from the destruction of the rain forest to the production of human insulin by genetically engineered bacteria, attempts to engage in this pursuit date back as far as the domestication of crops when metabiological evolution emerged, before the start of recorded history. Indeed, as we saw, Plato's *Republic* is a testimony to the attempts of society to nurture susceptible parts of growing minds to improve their "nature."

Perhaps the best metaphor of all for this process is Jonas Salk's description of his famous vaccine. Salk has described his vaccine as

an example of adding something unnatural (the chemical potentiator he added to the polio virus) to something natural (the virus itself,
inactivated) for the purpose of *improving upon nature* (that is, pre-
venting disease). This metaphor is well suited to the theory of knowledge and morality we have been discussing, wherein certain values
as well as the institutions that embody them have become understood as unnatural potentiators we add to the human side of nature
to improve upon nature—to get us into the collectively rational box
where human needs, wants, and interests are not so frustrated as in
the state of nature.

While Salk's added chemical "potentiates" the immune system, we
have seen how other kinds of potentiators nurture other systems. We
might thus say that Hume observed how we mix an unnatural moral
potentiator (justice) with the natural (social) world to try to improve
upon that natural world. Given the considerable extent to which we
literally construct reality (including tables, chairs, prize sheepdogs,
and human insulin for diabetic children), we can use this metaphor
as a new framework for understanding the relationships between
concepts, knowledge, values, and those social practices and institutions.

Once we set them within the Hegelian context of our entire story,
our resultant moral principles (our moral potentiators)—like our
knowledge acquisitions—can only be viewed as *achievements* that
result from our continuing search through our concepts and values
for the inconsistencies and inadequacies that reveal larger truths.
However, since as biological creatures we are only the beings we happen to be, with the particular needs, wants, and interests that have
arisen from evolution going the way it has happened to go, we must
not lose sight of this element of evolutionary history as we struggle
to apply our faculty of reason to obtain such noble achievements as
better facts and better values.

From Chapter 5 onward we have noted this historicism in any
philosophical theory that anchors itself to the actual biological
world, and it was Hegel who first reminded us how the shape of our
concepts can change the shape of world history. But from the first
pages of this book, we have seen how *ideas*—even some seemingly

academic ideas about nature and nurture—can have powerful practical consequences for psychiatric patients and other stigmatized
groups, for all school children, and for the shape of those social institutions that so influence human affairs at every level. The moral philosopher Alasdair MacIntyre has put it this way:

> Philosophy leaves everything as it is—except concepts. And since to
> possess a concept involves behaving or being able to behave in certain
> ways in certain circumstances, to alter concepts, whether by modifying
> existing concepts or by making new concepts available or by destroy-
> ing old ones, is to alter behavior. So the Athenians who condemned
> Socrates to death, the English Parliament which condemned Hobbes's
> *Leviathan* in 1666, and the Nazis who burned philosophical books
> were correct at least in their apprehension that philosophy can be sub-
> versive of established ways of behaving. Understanding the world of
> morality and changing it are far from incompatible tasks.

As we saw above, Salk has argued that our philosophy—the application of our minds—may now be called upon to take very seriously
the "directionality" of the term "better" facts and values as a new
historical era of teleobiological evolution emerges out of metabiological evolution. What would be the rules by which we should apply
our minds in searching concepts so powerful they could be threatening to the Athenians, the English Parliament, and the Nazi party?

Perhaps the most important rule is to remember how our moral
obligations must inevitably be tied to the historical question of what
we are at any given time possibly *capable* of achieving. A modern
view must accept that what it is possible to do sets limits on what we
are morally obliged to do. This was not always the case. During the
period of the Homeric epics, an admiral was no longer considered a
virtuous admiral even if he lost the battle only because of an unpre-
dictable freak storm that blew his ships away. Despite the ideal battle
plan and preparation of men and ships, the ancients assumed that if
he were virtuous, the gods would not have sent the storm. Today, we
would insist that what you can possibly do sets a constraint on the
domain of what it is virtuous to do.

In fact, this insistence in the secular domain that the "possibly can" limits the "moral ought" is relatively recent, dating back only two hundred years to Kant's definition: "Morality is the relation obtaining between action and the autonomy of the will." This Kantian idea has permeated the modern mind, and it is inseparable from our next rule: that *knowledge is responsibility*. This traditional notion is based on a complex view of free will. It basically says that if I know more, I can think more, and so in applying my faculty of reason I have more scope out of which to choose my actions. By expanding the range of what it is possible for me to do, I expand the domain of my moral obligations.

In our modern world, where the impossible often becomes possible in a matter of a few years—where we anticipate that the fraction of the human genome technologically capable of being "engineered" will go from almost nothing to the entire sequence in just a decade—the platitude that knowledge is responsibility takes on singular significance. After all, the human predicament that we have taken morality as a strategy to ameliorate is not at all mysterious. When we search our world for the ways in which human needs, wants, and interests are being frustrated—and frustrated in large measure—we can only agree with Hobbes that it is not scarce resources but scarce sympathies that are primarily the culprit. This "world vision" is itself in part a product of mass communication which could not have been possible when Hume contrasted his natural sympathy for a person he knew well as compared with "an Indian, or person wholly unknown to me." When that Indian's plight is brought into his home each evening on the television screen, our arguments about morality, based on species-defined characteristics, take on even greater significance.

Above, we took one example from Mackie, who suggested that patriotism might be a value which *was* useful but no longer *is* useful (even though it is at least as omnipresent across cultures as rectilinear architecture and Coca-Cola!). How can we evaluate this proposition? One further rule is that we must do so based on the *evidence*. If one result comes out of all the foregoing, it is that, based in a naturalistic biological view of the world, we must be "scientific" (in the broadest sense of that term) in our evaluation of our artificial virtues. We can look to local experiments to see what the results have been where different systems have been tried—and we can set up new experiments when this is feasible.

Hume's close friend Jean-Jacques Rousseau once wrote that "we are to take men as they are and moral laws as they might be." Perhaps this is not a bad starting point. Surely if even the rules of logic are ultimately held as valid on the pragmatic grounds of their empirical application, then the rules of ethics must be equally grounded in the realities of our biological life and social practices.

From this broadest of perspectives, as we saw with the breakdown of so many other of our time-honored distinctions in Chapters 13 and 14, even the distinction between nature and nurture begins to dissolve when we take a holistic enough view of nature that it must inevitably include in it whatever nurturing may be identified. We saw in that discussion how we might view even our DNA (the epitome of our "nature" side) as being a result of the "nurturing" of the environment in the Darwinian sense that environmental pressures have great bearing on which genes of a species survive into the next generation.

It is therefore tempting to break down one final distinction and remember that even Hume's "unnatural" virtues are *also* products of nature. The most artificial moral virtues we can contemplate are themselves the products of brain processes that have resulted from millions of years of "natural" selection, as our evolutionary ancestors interacted with a stressful environment and as species slowly adapted. Although we may say that evolution itself evolved and entered the metabiological era when human thoughts began to mix with it, those thoughts are in this sense as much products of biological evolution as was the appearance of new proteins that made our metabolism possible. Any meaningful distinction between nature and nurture has surely dissolved when we remember that *all* thought is input from what would have been called the nurture side, but thought itself arose as part of *nature*, which is the side usually taken to be thoughtless (as in Nietzsche's "indifferent beyond measure"). There is, however, one difference between mixing our thoughts with ongoing evolution and mixing some new metabolic protein with on-going evolution: the very notion of a teleobiological evolution presumes that we incur *responsibility* for our evolutionary experiments.

Taking the broadest possible perspective on nature, it is only natural for us to mix our thoughts with nature in our efforts to ameliorate the human condition. But this is not like any other natural process. Once we realize the extent to which we literally construct reality (our third direction), we also have a new challenge and responsibility to construct the *best* reality, applying our conceptual, moral, and social potentiators to improve upon nature—to nurture nature itself. It is no longer a question of *whether* we will engage in this process. It is only a matter of *how* we will engage in this process, and we must bring to bear on it all of the wisdom and humility we can muster as we take on this task of nurturing nature.

I mention both wisdom and humility because both are essential if we are not to make errors that could be truly disastrous. The two are inseparable. When the Delphic Oracle said of Socrates that he was the wisest man in Greece, Socrates' extensive discussions with many other men led him to conclude that the oracle realized that he alone knew that he did not know: the others thought they knew but did not. Salk has dubbed the post-Darwinian criterion for fitness in teleobiological evolution to be *wisdom*. When he speaks of the "survival of the wisest," he does not mean that wisdom will ensure the survival of particular individuals, but rather that wisdom will be required to ensure the survival of the species. To remember why humility is inseparable from this process, we need only look back at how those beautiful pure-bred dogs all end up with such bad hips.

But we also need to remember how anthropocentric the definition of morality is that has been adapted here from Hobbes, Warnock, and those game theorists who view all of ethics as an institution we humans create to improve our human predicament. As Thomas Benjamin, a friend of mine who studies the molecular genetics of viruses, is fond of reminding me: Salk's vaccine was not an improvement from the perspective of the polio virus! Most of the species that have ever lived have become extinct because they had the evolutionary equivalent of those bad hips, and nature does not think or care about whether this is a good thing or a bad thing, since nature is not in the business of thinking or caring about anything.

We are, however. The whole institution of ethics only arises with our thinking and caring about how things turn out, so—in keeping with the notion that knowledge is responsibility—we humans are the only moral agents in town. This is not to say that our ethics should only care about our own species. Other species also have some of the needs, wants, and interests that drive our considerations of values. But including nonhuman species in our human moral calculations is not the same as saying that other animals are moral agents—only that they are what the controversial contemporary ethicist Peter Singer has called "moral patients." (The lion that kills for its supper is not moral or immoral, just a lion being a lion.) Neither is this to say that we are able to make any ultimate comment on whether the disappearance of all ethics—because of the disappearance of all humans—would be "better" or "worse" in some objective, supra-moral, natural scheme of the world, since, again, *nature* does not care about this or anything else. It is only to say that, since even our intersubjective values are by definition "relative to us and the world as we and it happened to evolve," what we *mean* by ethics is inextricable from the human condition ethics was created to ameliorate. Simply put, our continued survival is valued in our ethical considerations because that is part of what we count as ethical considerations.

We end with the question of survival, because issues of survival have been fundamental to our descriptions of the human condition from both the individual and the collective perspective. At the individual level, Edelman showed us how even our basic cognitive concepts are constrained by values built into our brains, and these adaptive constraints (hunger, sex, and so on) all emerged from Darwinian evolution, which rewarded (with survival and reproductive success) those organisms that behaved in a way compatible with the maxim "survival and reproduction are better than no survival and reproduction." Indeed, with no other reference to our game-theoretic approach to ethics, Edelman concludes (based solely on the ways our

maps of maps can search for larger truths) that the process by which
our own ideas now mix with nature is not a process that operates
under Darwinian rules of nature. Using the phrase "informational
systems" in a broad context that includes the entire institution of
ethics, Edelman writes, "The contents of informational systems are
transferred by *use;* no genetic hereditary principle is needed." We
might call this process a product of nurture and not nature in the
sense that if we *stopped* using these informational systems, they
would quickly disappear—they would revert to wild-type like almost any artificial truth. Edelman recognizes the need to apply our
faculty of reason creatively to ameliorate the human condition, since
ethics has to do with the way the world ought to be (in the future),
and as a biologist he knows that the "present is not pregnant with a
fixed programmed future, and the program is not in our heads." He
thus hopes for a new Enlightenment, where what Salk has called
teleobiological evolution can solve the problems now confronting
humankind.

But this brings us to the question of survival at the collective level.
The age that has brought us mass communication (where we can
feel Hume's natural sympathy for that suffering person in India) has
also brought us weapons of mass destruction, in the forms of both
nuclear arsenals and environmental pollutants. For the first time in
history, we are facing global problems which potentially threaten the
collective survival of that species that gives rise to any meaning we
attach to any moral terms whatsoever. The question has therefore
been raised by many thinkers as to whether survival of the human
species is possible: metabiological evolution is relatively recent, and
teleobiological evolution has not itself had time to evolve very far.
Only a highly evolved teleobiological evolution could be relied upon
to solve such problems, which have never before presented themselves to the human mind.

There is no doubt considerable cause for concern here. Before his
death, the behavioral psychologist B. F. Skinner warned that we humans only learn by experience, and so it may not be possible to learn
to prevent the destruction of the species, since the events that would

teach us our mistakes might end any learning we will ever do as a species. When we move from moral values that attempt to ameliorate the predicament "my life is brutish, and it may become short" to new values that may be needed to address the possibility that "*all* life may become short," we enter a new era in the human institution we call ethics. Just as Hobbes based an entire theory of the state on considerations of individual survival, a completely new set of problems is raised by considerations of collective survival. In the language we have developed here, we might say that threats to collective survival require a fundamental change in the strategy morality has provided us for getting into the collectively rational box. After all, the strategy to date has been part of what game theory called a supergame, with Smith and Jones both knowing they will play round after round with the same rules. What happens when a new question arises and the outcome potentially makes this the last round of the game for all players?

In the theory of games, the appearance of the last round shifts the players from a supergame strategy to an endgame strategy. The incentives to keep promises or to honor treaties look very different when players believe they are in the final round. This is why, although life-boat ethics generally sheds little light on the ethical issues of the supergame here on the shore, endgame issues now permeate certain important global concerns. Indeed, endgame considerations have been a popular model for the proliferation of nuclear arms during the time when two superpower enemies were the principal players. Like Smith and Jones, both the United States and former Soviet Union could rationalize: "If they arm less, we are better off arming more, and if they arm more, we are better off arming even more." Even realizing intellectually that this put both countries in the individually rational box of mutual destruction, the possibility that this would be the last round made any psychological fetters too weak to move to the collectively rational box—and no world Leviathan exists to impose the equivalent of physical chains.

I am more optimistic than most about the power of human reason to apply itself to new problems. True, our reason has begun to reveal to us certain ways in which the traditional supergame morality of the metabiological evolutionary era is inadequate to solve new endgame problems. But this inadequacy has itself become one of Russell's metaphysical hooks that can grapple us to a new teleobiological evolutionary era in which we may be capable of learning before mistakes are made (one definition of wisdom). Edelman called this era a new Enlightenment based in part on a more sophisticated understanding of how our brains generate and process concepts and values. We might call it an era in which we apply new and better concepts and values, concepts and values artificially crafted by wise and humble minds for the specific task of nurturing our natural world.

We have already begun to see small successes in this process, where a new generation of young minds is being nurtured to be conscious of and sensitive to global environmental concerns and where international negotiations talk seriously about reducing the global nuclear threat. The artist, intellectual, and Czech President Vaclav Havel captured this spirit best when he accepted the American Liberty Medal in Philadelphia on July 4, 1994. Instead of celebrating the 218th anniversary of the Declaration of Independence, Havel delivered a new Declaration of *Inter*dependence. While Thomas Jefferson's individually rational declaration may have been appropriate to a liberal democracy in the world of the eighteenth century, Havel's new declaration captures the postmodern need for a collectively rational solution to the problems facing the world as we enter the new millennium—a collectively rational solution that requires a new vision of one interconnected human community. This is not to minimize the realistic causes for concern, given the complexity and gravity of the problems. It is, rather, a hint of the positive direction in which we must move if our facts and values are to keep pace with the evolving context of our evolving world.

The reminder above that perhaps the wisest man in history was wise because he knew he did not know should serve as an important warning when we begin to entertain the idea of teleobiological evolution. We should never forget the metaphor from the Indian subcontinent that compares our view of the universe to a small ant's view of a magnificent huge tapestry between a few of whose knotted strands the ant wanders its entire life. When we use the term teleobiological, we should not use it with some notion that we can chart evolution's course from some perspective outside of evolution that grasps all of evolution. Our abandonment of all external foundational solutions should make this obvious.

But if both our knowledge and our values are in an important sense "relative to" our human evolution, then there is an important *qualitative* difference between strategies of ethics that have heretofore sought to nurture individual minds to solve local prisoner's dilemmas and a new evolving strategy intentionally designed to solve the collective prisoner's dilemmas facing the species as a whole. As Salk puts it, the transition from metabiological to teleobiological evolution really begins when the evolution of consciousness reaches the point that *consciousness of evolution* becomes possible. If our artificial values are a device for ameliorating the human condition, then inculcating values for keeping promises and telling the truth constitutes a very different strategy from inculcating direct concerns for the continued flourishing of the species (the flourishing of which has been taken as the whole point of our moral enterprise). With global consciousness now beginning to arise, there may even be cause for cautious optimism that we can overcome that genetically determined tendency to care so much more about smaller than larger groups, and successfully nurture coming generations to think less in terms of culture, class, or country and more in terms of humankind.

We have come a long way from a simplistic debate about nature *versus* nurture. Having recognized that nurture is *a part of* nature (is emerging from it) and also *apart from* nature (in now trying to improve upon it), our synthetic view leaves us with a large graveyard of old distinctions and a new challenge to search for better facts and better values as we strive to improve the human condition and nurture that nature that has nurtured us.

Notes

Index

Notes

The references in the Preface are to C. Kramarae and P. A. Treichler, *A Feminist Dictionary* (London: Pandora, 1985), and to J. Piaget, "Comments on mathematical education," in A. G. Howson, ed., *Developments in Mathematical Education: Proceedings of the Second International Congress on Mathematical Education* (Cambridge: Cambridge University Press, 1973). The quotations that divide the parts of the book come from Aristotle, *The Nicomachean Ethics* (4th century B.C.), trans. J. A. K. Thomson (New York: Penguin, 1955); J. L. Austin, *How to Do Things with Words* (Oxford: Oxford University Press, 1955; revised by J. O. Urmson, 1980); G. M. Edelman, *Bright Air, Brilliant Fire* (New York: Basic Books, 1992); and H. L. A. Hart, *Law, Liberty, and Morality* (Stanford: Stanford University Press, 1963).

1. The State of the Art

As described in the Preface, this book develops and then extends to the realm of human values a theory of human knowledge that first appeared in E. M. Hundert, *Philosophy, Psychiatry, and Neuroscience: Three Approaches to the Mind* (Oxford: Oxford University Press, 1989). A reading of that first book is not assumed herein, and so the first half of this book and then Chapter 13 (which together develop the theory of knowledge) draw considerably from the earlier work, for which Oxford University Press is most gratefully acknowledged.

The original reference for Darwin is Charles Darwin, *On the Origin of Species by Means of Natural Selection or the Preservation of Favoured Races in the Struggle for Life* (New York: Appleton, 1859). The two Galton references are Francis Galton, *Hereditary Genius: An Inquiry into Its Laws and Consequences* (London: Macmillan, 1869), and Francis Galton, "The history of twins as a criterion of the relative powers of nature and nurture," *Royal Anthropological Institute of Great Britain and Ireland Journal* 6 (1876): 391–406. The story of the Jukes family and the nature-nurture "experiments" by Frederick II and emperor Akbar are described by G. A. Kimble in a chapter entitled "Evolution of the nature-nurture issue in the history of psychology" that opens the useful volume: R. Plomin and G. E. McClearn, eds., *Nature, Nurture, and Psychology* (Washington, DC: American Psychological Association, 1993).

The reference for Steven Pinker is *The Language Instinct: How the Mind Creates Language* (New York: William Morrow, 1994). The scientists who said that the genome sequence will teach us "how life works" and change "our philosophical understanding of ourselves" are James Dewey Watson and Walter Gilbert, respectively, in their separate contributions to D. J. Kevles and L. Hood, eds., *The Code of Codes: Scientific and Social Issues in the Human Genome Project* (Cambridge: Harvard University Press, 1992), from which the expression "code of codes" is taken. (The comparison of the Human Genome Project to the search for the Holy Grail comes from Walter Gilbert's contribution, entitled "A Vision of the Grail.") An excellent critical review entitled "The Dream of the Human Genome" was written by L. C. Lewontin in the *New York Review of Books*, vol. 39, no. 10 (May 28, 1992). The taboo against germ-line manipulation is most clearly articulated as one of the ten principles (commandments?) of "genethics" in D. Suzuki and P. Knudtson, *Genethics: The Ethics of Engineering Life* (Cambridge: Harvard University Press, 1990). A detailed presentation of the thalassemia screening program in Sardinia may be found in A. Cow, C. Rosatelli, R. Galanello, G. Monni, G. Olla, P. Cossu, and M. S. Ristaldi, "The prevalence of thalassemia in Sardinia," *Clinical Genetics* 36 (1989): 277–285.

2. The Divide-and-Conquer Strategy

For a most accessible account of Plato's philosophy, the reader is recommended to R. M. Hare, *Plato* (Oxford: Oxford University Press, 1982), in which Hare makes the point about our skepticism of the sinner who claims he is a saint "but for the sin that dwelleth within me." All the Plato quotations may be found in Hare's *Plato*, which used Stephanus's edition as printed in the margin of the standard *Oxford Classical Text of Plato* (ed. J. Burnet, 1900–1907). The quotation spoken by Meno is from the dialogue *Meno*, and the reference to reason's ability to prevail when a thirsty person avoids drinking impure water comes from the *Phaedrus*, although Plato's divide-and-conquer model of the mind is even more fully described in the *Protagoras* and the *Republic*. The discussion of Aristotle's views comes from *The Nicomachean Ethics*, referenced above.

The best overview of Sigmund Freud's model of the mind is probably his own in *The Ego and the Id* (1923), which is contained in volume 19 of *The Standard Edition of the Complete Psychological Works of Sigmund Freud* (London: Hogarth Press and the Institute of Psycho-Analysis, 1953–1974). Freud's entire theory is summarized by him in his *New Introductory Lectures on Psycho-Analysis* (1933) in volume 22 of the *Standard Edition*, where his metaphor of a small man (the ego) riding a large horse (the id) may be found.

The quotation from David Hume appears in *A Treatise of Human Nature* (1740; Oxford: Oxford University Press, 1978), and the reference to Kant alludes to I. Kant, *Critique of Practical Reason*, trans. L. W. Beck (1788; Chicago: University of Chicago Press, 1949). The discussion of the radically different Hopi experience of time as a "hoax" may be found in Pinker, *The Language Instinct*, referenced above.

3. Constructing Experience

Kant's observation that it "seems at first strange" to think that the understanding prescribes the laws of nature comes from *Prolegomena to Any Future Metaphysics*, trans. P. Carus and J. W. Ellington (1783; Indianapolis: Hackett Publishing Company, 1977), as does the quotation that the laws of nature "are not derived from experience, but experience is derived from them." His famous statement that "thoughts without content are empty" comes from *Critique of Pure Reason*, trans. J. M. D. Meikeljohn (1781; London: J. M. Dent & Sons, 1934). The reader is recommended to P. F. Strawson, *The Bounds of Sense* (London: Methuen & Company, 1966), for a detailed discussion of the form-content distinction and Kant's general theory. Hume's theory is outlined in *A Treatise of Human Nature*, referenced above. The reference for John Locke is *An Essay Concerning Human Understanding* (1690; Oxford: Clarendon Press, 1975). The reference for Sir Isaac Newton is *Principia*, trans. A. Motte, revised by F. Cajori (1686; Berkeley: University of California Press, 1934).

It is worth noting that Kant actually distinguished the dimensional qualities of time and space as what he called "pure forms of sensible intuitions," in contrast to "categories of the faculty of understanding." By calling these forms of sensibility "pure," he meant to imply "prior to experience" in a similar way that his a priori categories are "prior to experience." This was important for Kant since he built from this notion a number of (long-since discarded) propositions about the philosophical status of geometry and mathematics in general. For our purposes, Kant's distinction between the pure and the a priori is unimportant, so the term "category" is used to refer to all a priori concepts, including space and time. For a discussion of why this part of Kant's distinction fails to hold up on detailed analysis, see Strawson, *The Bounds of Sense*, or Hundert, *Philosophy, Psychiatry, and Neuroscience: Three Approaches to the Mind*, both referenced above.

4. Thoughts and Things

The title of this chapter is a tribute to Frederick Will's inspirational address, "Thoughts and things," *Proc. and Addresses Amer. Phil. Assn.* 42 (1968–1969): 51–69, which inspired a great deal of the thought in this book. Will's lovely comment about the existence of dogs and cats comes from *Induction and Justification: An Investigation of Cartesian Procedure in the Philosophy of Knowledge* (Ithaca, NY: Cornell University Press, 1974). The remainder of the quotes from Will all come from "Thoughts and things," referenced above. Descartes's method of doubt is presented in his *Meditations* (1641), which may be found in *René Descartes: The Essential Writings*, trans. and ed. J. J. Blom (New York: Harper & Row, 1977). The best discussion of the active-passive distinction in Descartes and Kant may be found in J. F. Bennett, *Kant's Dialectic* (Cambridge: Cambridge University Press, 1974).

The analysis of Hegel's thought comes largely thanks to the inspiration of R. Solomon, *In the Spirit of Hegel* (Oxford: Oxford University Press, 1983), wherein Hegel's own description of his *Phenomenology* is described. The bud/blossom/fruit metaphor comes directly from G. W. F. Hegel, *The Phenomenology of Spirit*, trans. A. V. Miller (1807; Oxford: Oxford University Press, 1977). The comparative anatomist metaphor may be found in B. Russell, *The Problems of Philosophy* (1912; Oxford: Oxford University Press, 1967).

5. How Babies Think Things Over

The quotation "how we think with the help of things" comes from Will's "Thoughts and things," referenced above. For Piaget's comprehensive theory of how genetic principles might relate objects and concepts, see J. Piaget, *The Principles of Genetic Epistemology*, trans W. Mays (1970; London: Routledge & Kegan Paul, 1972).

6. How Quickly They Grow Up

The direct reference is to J. Piaget, *The Origins of Intelligence in Children*, trans. M. Cook (1936; London: Routledge & Kegan Paul, 1952). However, the overview of Piaget's theory comes equally from J. Piaget, *The Construction of Reality in the Child*, trans. M. Cook (1937; New York: Basic Books, 1954), as a primary source, and from J. H. Flavell, *The Developmental Psychology of Jean Piaget* (New York: D. Van Nostrand Co., 1963), as a most useful secondary source. For a good, short, basic introduction to Piaget's life and work, see M. A. Boden, *Piaget* (London: Fontana Paperbacks, 1979), which was useful in preparing the brief overview in this chapter.

David Elkind is the student of Piaget whose definition of genetic epistemology is provided, and this comes from his "Editor's introduction" to J. Piaget, *Six Psychological Studies* (Sussex: The Harvester Press, 1980). The quotation about how the relationship between thoughts and things is really part of the relationship of an organism to its environment may be found in J. Piaget, *The Child's Conception of Physical Causality*, trans. M. Gabain (1927; London: Routledge & Kegan Paul, 1930).

Freud's stage theory is summarized in his *New Introductory Lectures on Psycho-Analysis*, referenced above, and Erikson's may be found in *The Life Cycle Completed* (New York: Norton, 1982). A wonderful summary of L. Kohlberg's stages of moral development is found in "From is to ought: how to commit the naturalistic fallacy and get away with it in the study of moral development," in T. Mischel, ed., *Cognitive Development and Epistemology* (New York: Academic Press, 1971). The contrasting view is best found in C. Gilligan, *In a Different Voice* (Cambridge: Harvard University Press, 1982).

7. Feelings and Things

All of the quotations from Arnold Modell come from *Object Love and Reality* (New York: International Universities Press, 1968), to which much of this chapter and the entire book are indebted. The notion of identifying with the feelings of a baby in the womb may be found in E. H. Erikson, *Identity: Youth and Crisis* (New York: Norton, 1968). D. W. Winnicott's views on early development are described in *Playing and Reality* (New York: Basic Books, 1971). The first reference to his concept of the good enough mother may be found in his earlier *Collected Papers* (New York: Basic Books, 1958).

Kegan's reminder about the evolutionary message of some of the baby's innate reflexes is found in *The Evolving Self* (Cambridge: Harvard University Press, 1982). The researchers extending Piaget's methodology to study the development of basic concepts in the mentally ill are S. J. Blatt and C. M. Wild, and the quotation comes from their *Schizophrenia: A Developmental Analysis* (New York: Academic Press, 1976). Bell's study was first published in S. M. Bell, "The development of the concept of object as related to infant-mother attachment," *Child Development* 41 (1970): 292–311. I have discussed the possibility that psychotic disorders can be usefully understood as distortions of Kantian categories in E. M. Hundert, "Are psychotic illnesses category disorders?" in M. Spitzer and B. Maher, eds., *Philosophy and Psychopathology* (New York: Springer-Verlag, 1990). The best description of Piaget's account of the infant's early experience of pictures which disappear and reappear capriciously is found in his *The Construction of Reality in the Child*, referenced above.

The studies of infant development in the absence of nurturing that are alluded to at the end of the chapter are described in R. A. Spitz, *The First Year of Life* (New York: International Universities Press, 1965); J. Bowlby, *Attachment and Loss*, 2 vols. (New York: Basic Books, 1969 and 1973); and A. Freud, *The Ego and the Mechanisms of Defense* (1936; rev. ed. New York: International Universities Press, 1966).

An unfortunate confusion exists in the psychoanalytic jargon between Piaget's cognitive concept of "object permanence" and the term "object constancy," originally introduced by H. Hartmann in *Ego Psychology and the Problem of Adaptation* (1939; New York: International Universities Press, 1958). Generally speaking, object constancy is used to refer to an affectively charged ability to maintain person-permanence with respect to mother when she is gone. Since a "person" is a subset of all "objects" whose permanence the child can assimilate, it is tempting to claim on logical grounds that object permanence is a prerequisite for object constancy in infant development.

The situation is actually more complex, however, since children go through an intermediate period during the middle of the sensorimotor stage where seeing *part* of an object can remind them of the whole object— this recognition memory being simpler than evocative memory where the image must be evoked without any stimulus. (This is why students prefer multiple-choice tests to a fill-in-the-blanks format.) When it comes to the situation in the nursery, those *internal* feelings of being cold, wet, and hungry can serve as the stimuli for remembering that mother should be on her way to help, and this affective stimulus is a powerful connection to the developing concept of person- (that is, mother-) permanence. In this context, we should not be surprised by Bell's finding that a higher quality of infant-mother attachment can cause object constancy to precede and herald the development of the concept of object permanence of affectively neutral "things." It likewise should come as no surprise that when children play that pick-the-row-with-more-stones game (Chapter 6), they manage to choose the larger number correctly months earlier when playing the game with chocolates rather than stones.

8. The Anthropic Principle

The Hegel quotation comes from his *Phenomenology of Spirit*, although the Hegelian analysis is again indebted to Solomon's *In the Spirit of Hegel*, both referenced above. J. L. Austin's ideas about performative language are found in *How to Do Things with Words*, also referenced above. The reference for S. Hawking is *A Brief History of Time* (Bantam Press, 1988). A most complete review of the anthropic principle may be found in J. D. Barrow and F. J. Tipler, eds., *The Anthropic Cosmological Principle* (Oxford: Oxford University Press, 1986). The Kant quotation is again from his *Prolegomena to any Future Metaphysics*, referenced above. M. Heidegger's approach may be found in *Being and Time* (1927; New York: Harper & Row, 1972).

9. The Facts of Life

Spemann's original experiments on the determination and differentiation of embryonic neurons are described in *Embryonic Development and Induction* (New Haven: Yale University Press, 1938). Details of the infant's reflex to turn its eyes toward specific features of high contrast may be found in B. M. Wilcox, "Visual preferences of human infants for representations of the human face," *J. Exper. Child Psychol.* 7 (1969): 10–20, and the original study of the patterns of stimuli that will elicit a smile from a six-week-old baby was R. Ahrens, "Beitrage zur Entwicklung des Physiognomic- und Mimiker-kennens," *Z. Exp. Angew. Psychol.* 2 (1954): 412–454, 599–633. The allusion to Darwin's observations draw on *The Expression of the Emotions in Man and Animals* (London: Murray, 1872).

10. The Perceptive Brain

For a detailed account of how the brain evolved "from the inside out" see H. B. Sarnat and M. G. Netsky, *Evolution of the Nervous System* (New York: Oxford University Press, 1981). The division of cortical regions by microscopic anatomy of large and small cell layers was described in C. F. von Economo, *The Cytoarchitechtonics of the Human Cortex* (London: Oxford Medical Publications, 1929). Other general anatomical points may be found in M. Carpenter and J. Sutin, *Human Neuroanatomy* (Baltimore: Williams and Wilkins, 1983), or any other standard text. The physiological points may be found in E. R. Kandel, J. H. Schwartz, and T. M. Jessell, *Principles of Neural Science*, 3rd edition (Norwalk, CT: Appleton and Langer, 1991). The reference to Oliver Sacks is to *The Man Who Mistook His Wife for a Hat* (New York: Summit Books, 1985; Harper & Row, 1987).

The division of the mind into input, central processing, and output systems was also made on theoretical grounds in J. Fodor, *The Modularity of Mind* (Cambridge: The MIT Press, 1983), as described in some detail in my earlier *Philosophy, Psychiatry, and Neuroscience: Three Approaches to the Mind*, referenced above. The distinct pathways for object vision and spatial vision were detailed in M. Mishkin, L. G. Ungerleider, and K. A. Macko,

"Object vision and spatial vision: two cortical pathways," *Trends NeuroSci.*
6 (1983): 414–417, and the syndrome of hemispatial neglect is described in
K. M. Heilman and E. Valenstein, *Clinical Neuropsychology* (Oxford: Oxford
University Press, 1985). The best original source for the columnar structure
of the visual cortex is the classic D. H. Hubel and T. N. Wiesel, "Functional
architecture of macaque monkey visual cortex," *Proc. R. Soc. London* (Ser.
B) 198 (1977): 1–59. A nice overview of the role of NMDA receptors may
be found in G. L. Collingridge and T. V. P. Bliss, "NMDA receptors: their
role in long-term potentiation," *Trends NeuroSci.* 10: 288–293.

As scientists uncover the detailed workings of different subregions of the
association cortex, some have even suggested that we abandon the term
"association cortex," which was historically used to label those areas whose
functions we have not yet identified. I have preserved its use here because
even if the function of every single cell were understood, the impressive fact
remains that cortical regions with homotypical histology (large and small
cells scattered among the six cell layers) synthesize information from multiple sensorimotor building blocks, in contrast with the primary input and
output areas whose granular and agranular histology, respectively, is concerned with the angular orientation of visual input, the movement of one
finger, or the analysis of some particular aspect of language. In synthesizing
these building blocks together, our homotypical brain regions may still collectively be called association areas, even when the specific ways they do
their associating have all been understood.

11. The Plastic Brain

The earlier insights about the effects of newborn cataracts are described in
M. von Senden, *Space and Sight*, trans. P. Heath (Glencoe, IL: Free Press,
1960). Riesen's work with monkeys raised in complete darkness is detailed
in A. H. Riesen, "Plasticity of behavior: psychological aspects," in H. F. Harlow and C. N. Wolsey, eds., *Biological and Biochemical Bases of Behavior*
(Madison: University of Wisconsin Press, 1958). Harry F. Harlow's experiments on social isolation are described in H. F. Harlow, R. O. Dodsworth,
and M. K. Harlow, "Total social isolation in monkeys," *Proc. Natl. Acad. Sci.
USA* 54 (1965): 90–97. The examples of infant language plasticity in the
discrimination of specific sounds come from Pinker, *The Language Instinct*,
referenced above. The universality of infant babbling is discussed in E. Len-
neberg, *Biological Foundations of Language* (New York: Wiley, 1967). Spitz's
description of anaclitic depression in institutionalized human infants is
found in "Anaclitic depression," *Psychoanalytic Study of the Child* 2 (1946):
313–342.

The best overview of the discovery of neuroplasticity is probably Wiesel's

Nobel lecture, reprinted in T. N. Wiesel, "The postnatal development of the visual cortex and the influence of the environment," *Bioscience Reports* 2 (1982): 351–377, from which the quotations are taken, as are the estimates for the duration of visual critical periods in kittens, macaques, and humans. The example given of the extreme degree of plasticity was reported in C. Blakemore and D. E. Mitchell, "Environmental modification of the visual cortex and the neural basis of learning and memory," *Nature* 241 (1973): 467–468. The recognition of the important roles of GAP-43 was described in L. I. Benowitz and A. Routtenberg, "A membrane phosphoprotein associated with neural development, axonal regeneration, phospholipid metabolism, and synaptic plasticity," *Trends NeuroSci.* 10 (1987): 527–532. The distribution of GAP-43 in the adult human brain is graphed in M. Mishkin and T. Appenzeller, "The anatomy of memory," *Scientific American* 256 (1987): 80–89. (Routtenberg's "F-1 protein" is another name for GAP-43, as described in Benowitz and Routtenberg, just above.) Recent studies proving that GAP-43 is a critical determinant of cell growth are found in L. Aigner and P. Caroni, "Depletion of 43-kD growth-associated protein in primary sensory neurons leads to diminished formation and spreading of growth cones," *Journal of Cell Biology* 123 (1993): 417–429.

Although the GAP-43 (pre-synaptic) side of the visual plasticity story is highlighted rather than the NMDA (post-synaptic) side, excellent reviews of the latter include K. D. Miller, B. Chapman, and M. P. Stryker, "Visual responses in adult cat visual cortex depend on N-methyl-D-aspartate receptors," *Proc. Natl. Acad. Sci. USA* 86 (1989): 5183–5187; M. F. Bear, A. Kleinschmidt, Q. Gu, and W. Singer, "Disruption of experience-dependent synaptic modifications in striate cortex by infusion of an NMDA receptor antagonist," *Journal of Neuroscience* 10 (1990): 909–925; and K. Fox and N. W. Daw, "Do NMDA receptors have a critical function in visual cortical plasticity?" *Trends NeuroSci.* 16 (1993): 116–122.

It is also worth noting that plasticity can occur in the nervous system not only through anatomical changes, as described in this chapter (how orientation columns hook themselves up), but also through functional changes, where the same physical connections can become used for new functions that are in increasing demand. Thus, the neurophysiologist M. M. Merzenich has shown that the somatosensory cortical map of the "touch" inputs from different regions of an adult owl monkey's hand can undergo a reversible *functional* shift as when, say, the area representing inputs from one finger takes over much of the map of the entire hand when a monkey is trained to continually tap that finger. (See W. M. Jenkins, M. M. Merzenich, M. T. Ochs, T. Allard, and E. Guic-Robles, "Functional reorganization of primary somatosensory cortex in adult owl monkeys after behaviorally controlled tactile stimulation," *Journal of Neurophysiology* 63 (1990): 82–

104.) Although even this distinction between anatomical and functional changes is itself yet another blurry distinction (at the electron microscope level, the *molecular* anatomy must be changing as synapses strengthen and weaken), it is presumably this kind of functional plasticity that accounts for the continued ability of the senses to be "educated" after the critical period. As William James pointed out (in defense of empiricism) in *The Principles of Psychology* (New York: Holt, 1890), we must not forget the "well-known virtuosity displayed by the professional buyers and testers of various kinds of goods. One man will distinguish by taste between the upper and lower half of a bottle of old Madeira. Another will recognize, by feeling the flour in a barrel, whether the wheat was grown in Iowa or Tennessee."

12. Even Robots Need Values

Edelman's first description of his theory of neuronal group selection was published as the chapter "Group selection and phasic re-entrant signaling: a theory of higher brain function" in G. M. Edelman and V. B. Mountcastle, *The Mindful Brain* (Cambridge: The MIT Press, 1978). After several further books on the theory, his most recent and readable account is *Bright Air, Brilliant Fire* (referenced above), from which all the Edelman quotations are taken. A recent excellent paper from Edelman's group which demonstrates how neural systems "categorize on value" is K. J. Friston, G. Tonini, G. N. Reeke Jr., O. Sporns, and G. M. Edelman, "Value-dependent selection in the brain: Simulation in a synthetic neural model," *Neuroscience* 59 (1994): 229–243.

Hebb's original findings are discussed in his now classic *The Organization of Behavior: A Neuropsychological Theory* (New York: Wiley, 1949). I first heard the reframing of Hebb's observations in the terms "neurons that fire together wire together" in a lecture by the neurobiologist Carla J. Schatz entitled "Order from disorder in visual system development," which she presented at the Marine Biological Laboratory at Woods Hole on July 29, 1994. Professor Schatz has demonstrated that activity-dependent competition determines the shape of the visual system even during gestation, when spontaneous firing of retinal ganglion cells is used to create the precise connections found in the lateral geniculates even before the system "sees" anything (since this all happens in the darkness of the uterus).

The view that children can learn language because they already can make some sense of the world is discussed in M. Donaldson, *Children's Minds* (New York: Norton, 1978). The reference for Lakoff's proposal about how language is based on cognition (rather than vice versa) is *Women, Fire, and Dangerous Things: What Categories Reveal about the Mind* (Chicago: University of Chicago Press, 1987).

A good example of the more complicated psychiatric understanding of how a mental illness might be caused by a biochemical imbalance *not* completely determined by genetics may be found in D. R. Weinberger, "Implications of normal brain development for the pathogenesis of schizophrenia," *Arch. Gen. Psychiatry* 44 (1987): 660–669. The quotations at the end of the chapter are from Modell, *Object Love and Reality,* and Blatt and Wild, *Schizophrenia: A Developmental Analysis,* both referenced above.

13. But How Can You Be Sure of That?

This chapter is the one most directly taken from my earlier *Philosophy, Psychiatry, and Neuroscience: Three Approaches to the Mind,* referenced above, particularly subsections 3 and 4 from Chapter 2 of that book. The concept was originally suggested to me by Thomas P. Steindler in an unpublished 1978 essay entitled "Conceptual boundaries." Kuhn's description of how we can use the science we have to shake the foundations of that science is found in T. S. Kuhn, *The Structure of Scientific Revolutions* (Chicago: University of Chicago Press, 1962). Ludwig Wittgenstein's metaphor of the sandy riverbed is from *On Certainty,* trans. D. Paul and G. E. M. Anscombe (1950; New York: Harper & Row, 1972).

The question about what supports logic as logic does mathematics, and so on, was put forth in R. S. Root-Bernstein, *Discovering* (Cambridge: Harvard University Press, 1989). The quotation about the pragmatism of logic is from C. I. Lewis, "A pragmatic conception of the *a priori*," which was originally published in *The Journal of Philosophy* in 1923 and reprinted in H. Feigl and W. Sellars, eds., *Readings in Philosophical Analysis* (New York: Appleton-Century-Crofts, 1949). The first reference I know of to the expression "biological necessity" in the context of this kind of discussion is Noam Chomsky's in *Reflections on Language* (New York: Pantheon Books, 1975).

The distinction between coherence and correspondence criteria for truth (and also between a criterion of truth and a theory of truth) loom large in analytical philosophy, but are not important for the naturalized epistemology presented here. For a penetrating discussion of these distinctions and related points, see K. R. Westphal, *Hegel's Epistemological Realism* (Dordrecht: Kluwer, 1989).

14. Breaking Down Old Distinctions

For outstanding recent books by Putnam and Quine, see H. Putnam, *Realism with a Human Face* (Cambridge: Harvard University Press, 1990), and W. V. Quine, *Pursuit of Truth,* rev. ed. (Cambridge: Harvard University

Press, 1990). The idea that science is the search for "things worth naming" is from C. I. Lewis, "A pragmatic conception of the *a priori*," referenced above. Howard Gardner's observation about the new, "natural" view of concepts is from *The Mind's New Science* (New York: Basic Books, 1985). Russell's metaphor of Hegelian "metaphysical hooks" is from *The Problems of Philosophy,* referenced above. The quotation about the areas of rectangles comes from G. A. Kimble, "Evolution of the nature-nurture issue in the history of psychology," also referenced above.

15. Everything Is Relative?

Piaget's belief that the structures he studied "hold true for the species" is a quotation from Elkind's "Editor's introduction" to J. Piaget, *Six Psychological Studies,* referenced above. Rosch's original cross-cultural color-naming studies are described in E. Rosch, "Natural categories," *Cognitive Psychol.* 4 (1973): 328–350, and E. Rosch, "Cognitive reference points," *Cognitive Psychol.* 7 (1975): 532–547. The quotation from Pinker comes from *The Language Instinct,* referenced above. The order in which primitive cultures assign color names may be found in B. Berlin and P. Kay, *Basic Color Terms: Their Universality and Evolution* (Berkeley: University of California Press, 1969). The two quotations from Wiggins come from D. Wiggins, "Truth, invention, and the meaning of life," *Proc. Brit. Academy* 66 (1976): 331–378. The Edelman quotation is again from *Bright Air, Brilliant Fire,* referenced above.

It is worth pointing out that John Locke made his famous distinction between primary qualities like object-ness (which he took to be properties things "really" have) from secondary qualities like color (which he took to arise only from our interactions with things) based on the then-recently energized science of physics. With the birth of optics, Locke saw that some physical *texture* (a primary quality) at the surface of objects is a property they really *have,* color being a secondary property we attribute to objects as this "surface texture" determines which wavelengths of light are absorbed or reflected to our eyes. As neurobiology replaces physics as the scientific paradigm for philosophy, the old distinction between primary and secondary qualities becomes less central, and the new distinction between qualities handled separately by our faculties of sensibility and understanding (the former with and the latter without critical periods) become matters for empirical discovery.

In *Philosophy, Psychiatry, and Neuroscience: Three Approaches to the Mind,* I suggested the notation "modules" and "categories" for these qualities, respectively, with the caveat that the distinction is as blurry as Locke's turned out to be, and with a reminder that the shift from physics to neuroscience is part of what moves us from a Lockean correspondence theory of truth to our coherence theory of truth. After all, when it is determined that temperature is nothing but motion (the motion of small particles), Locke must decide whether to take a primary quality off his list, since it no longer corresponds to a property objects *really have*. Neuroscientists would not change their list of perceptual modules, however: temperature may *be* motion on a different scale, but our input systems treat motion and temperature as totally different kinds of input.

16. Inventing Right and Wrong

This chapter grew out of the appendix (entitled "Cognitive and moral intersubjectivity") of my *Philosophy, Psychiatry, and Neuroscience: Three Approaches to the Mind*, referenced above. The best discussion of the status of "objective values" that I have found is the first section of J. L. Mackie, *Ethics—Inventing Right and Wrong* (Penguin Books, 1977), to which the title of this chapter is a tribute. In that same work, Mackie also makes the points that (a) when evolution replaces an omniscient God in the unfolding of history, then we are at best left with values that *were* adaptive (his patriotism example), and (b) that values with larger reference groups are less specific in how they guide action but are more stable over time. The wonderful synthesis of Aristotle's "non-cognitivist" and Spinoza's "cognitivist" views that is framed in terms of "two equally valid 'becauses'" comes from D. Wiggins, "Truth, invention, and the meaning of life," referenced above. The discussion of "social facts" (who has the power in society?) comes from S. Lukes and W. G. Runciman, "Relativism: cognitive and moral," *Proc. Arist. Soc.* (supplementary vol.) 48 (1974): 165–208. The book summarizing our species-defined needs, wants, and interests is D. E. Brown, *Human Universals* (New York: McGraw-Hill, 1991).

17. The Prisoner's Dilemma

The leap from "is" to "ought" comes from D. Hume, *A Treatise of Human Nature*, referenced above. Mackie's discussions of morality as an institution and moral reasoning as a species of game theory come from his *Ethics—Inventing Right and Wrong*, also referenced above. G. J. Warnock's argument may be found in *The Object of Morality* (London: Methuen & Co., Ltd., 1971). The quotation from Hart comes from his *Law, Liberty, and Morality*, referenced above. The suggestion that Aristotle's famous comment should be taken as a biological observation comes from J. Barnes, *Aristotle* (Oxford: Oxford University Press, 1982). Tocqueville's comment comes from A. de Tocqueville, *Democracy in America* (1839; New York: Vintage Classics,

1990). The Dawkins reference is R. Dawkins, *The Selfish Gene* (New York: Oxford University Press, 1976).

The remaining references in political economy are J. S. Mill, *Principles of Political Economy* (1848; New York: D. Appleton, 1884); K. Marx, *Capital: A Critique of Political Economy,* 3 vols. (1867, 1893, 1894; New York, International Publishers, 1967); T. Hobbes, *The Leviathan* (1651; London: J. M. Dent, 1914); and J. M. Keynes, *General Theory of Employment, Interest, and Money* (London: Macmillan, 1936).

Although Warnock's notion of an "object of morality" takes morality to be an "institution" in a very broad sense of that term, Mackie has argued (in *Ethics: Inventing Right and Wrong,* referenced above) that this same theory takes "morality" in a relatively narrow sense of that term. Mackie distinguishes morality in the broad sense as the guiding principles that govern all of human behavior from morality in the narrow sense as the limited principles that deal with particular kinds of tradeoffs between competing values. In the narrow but not the broad sense, it thus becomes possible to say: "I know morality dictates X, but there are other considerations here—say, cultural or aesthetic ones—that dictate Y instead." Although he claims Warnock's theory limits itself to morality in the narrow sense, the synthetic view presented here must ultimately break down distinctions between ethical, cultural, aesthetic, and other such considerations, blurring this version of a distinction between ethics in the narrow and broad sense.

But Mackie also makes another, closely related distinction that remains very important for the theory presented here: the distinction between first-order, practical ethical theories that concern human behavior and second-order, metaethical theories in moral philosophy. It is worth noting how the metaethical idea that morality has an object or a point to it differs from traditional first-order teleological theories of ethics that define the "good" in terms of getting us closer to some *telos* (some goal, like the greatest happiness for the greatest number in utilitarian theory). By distinguishing first-order moral theories that deal with how we humans ought to behave from second-order metaethical theories about the philosophical status of first-order theories, Mackie shows how a theory like the one presented here might be considered teleological only in its second- but not necessarily first-order position. Ethics only *exists* to help ameliorate the human condition, but this need not commit us to a teleological set of first-order moral principles. It may well be that the best way for morality to achieve its second-order meta-objective is to develop a first-order moral theory that consists entirely of some set of deontological rights, shalt's, and shalt not's that could not be more different from consequentialist moral maxims.

18. Natural Law, Nurtural Law

Hume's discussion of "natural versus artificial virtues" may be found in his *A Treatise of Human Nature*, referenced above. A. MacIntyre's comments about the "objective" force of religious moral language come from *A Short History of Ethics* (New York: Macmillan, 1966), and is further elaborated in *After Virtue: A Study in Moral Theory* (London: Gerald Duckworth, 1981). The Dalai Lama's approach to transforming the mind through compassion is outlined in T. Gyatso, *Kindness, Clarity, and Insight*, trans. J. Hopkins (Ithaca, NY: Snow Lion Publications, 1984). Kant's analysis of "universalizability" is most famously found in I. Kant, *Critique of Practical Reason*, trans. L. W. Beck (1788; Chicago: University of Chicago Press). The idea that it may be natural to want to be the most virtuous member of the tribe (even though the particular definition of virtue will always be artificial) was suggested to me in a conversation with Jerome Kagan at Harvard.

The argument about rationality and evolution and the example of the thief performing an irrational but adaptive syllogism is found in B. Maher, "The irrelevance of rationality to adaptive behavior," in Spitzer and Maher, eds., *Philosophy and Psychopathology*, referenced above. Edelman's estimate of the date of appearance in evolution of primitive consciousness comes from his *Bright Air, Brilliant Fire*, referenced above. E. O. Wilson's still classic framework may be found in *Sociobiology: A New Synthesis* (Cambridge: Harvard University Press, 1975). J. Salk's framework is put forth in *The Anatomy of Reality* (New York: Praeger Publishers, 1985).

19. Lessons from an Optical Illusion

R. L. Gregory's account of the Müller-Lyer illusion is found in *Eye and Brain: The Psychology of Seeing* (London: Weidenfeld and Nicolson, 1966). J. B. Deregowsky's studies of the Zulus' perception of the illusion are described in "Illusion and culture," in R. L. Gregory and E. H. Gombrich, eds., *Illusion in Nature and Art* (New York: Scribner, 1974). J. L. Austin's wonderful quotation about over-simplification in philosophy comes from *How to Do Things with Words*, referenced above. A penetrating discussion of Hegel's use of the notion of *Sittlichkeit* may be found in R. Solomon, *In the Spirit of Hegel*, referenced above. J. Rawls's model of a "reflective equilibrium" was put forth in *A Theory of Justice* (Cambridge: Harvard University Press, 1971). I have discussed the relationship between small-reference-group concepts and psychotic delusions in "Are psychotic illnesses category disorders?" referenced above. The moral tenets of the organized psychiatric profession are codified in *The Principles of Medical Ethics with Annotations*

Especially Applicable to Psychiatry, published and updated annually by the American Psychiatric Association, Washington, D.C.

20. Nurture Improving upon Nature

The argument about return to wild-type was suggested by Brendan and Barbara Maher in a series of stimulating conversations. The quotations from Friedrich Nietzsche come from *Beyond Good and Evil,* trans. R. J. Hollingdale (1886; New York: Penguin Books, 1990). The metaphors from Salk come from *The Anatomy of Reality,* referenced above, and from *Survival of the Wisest* (New York: Harper & Row, 1973). The MacIntyre quotation about the role of philosophy in history comes from the opening pages of *A Short History of Ethics,* referenced above. The analysis of Hume's "Indian, or person wholly unknown to me" comes from J. L. Mackie, *Ethics—Inventing Right and Wrong,* also referenced above. The Rousseau quotation comes from *The Social Contract,* trans. C. M. Cherover (1762; New York: Harper & Row, 1984). The quotation from Kant comes from *Critique of Practical Reason,* referenced above. Peter Singer's distinction between "moral agents" and "moral patients" comes from *Practical Ethics* (Cambridge: Cambridge University Press, 1979). Edelman's quotations and his notion of a "new Enlightenment" come from *Bright Air, Brilliant Fire,* also referenced above. B. F. Skinner's later thought is presented in *Upon Further Reflection* (Englewood Cliffs, NJ: Prentice-Hall, 1987).

In this final chapter, I make the book's only allusion to the notion of free will, recognizing that some further discussion of this subject might be expected. I apologize for not adding this huge question to the tasks of this book with the following rationale: Either free will exists as we intuit, or everything really is determined. If determinism is true, then every time we discuss the question of free will versus determinism, we have no choice but to do so. To me, this presents an obvious strategy—any time we even suspect we might be free to avoid repeating this discussion, we should seize the opportunity.

Index

Accommodation, 68, 69, 72, 73, 121, 151; to environment, 73, 114, 132, 156, 181; cellular, 124–125, 181–182; of concepts/conceptual, 145, 151–152. *See also* Adaptation; Assimilation

Adaptation, 73, 99, 137, 181; to environment, 100, 122, 185; in evolution, 176, 182–183, 204; constraints on, 228. *See also* Accommodation; Assimilation

Aesthetics, 51, 201, 212–213, 214, 217

Akbar, 5, 236

Altruism, 192–193

Analytic vs. synthetic, 153–154, 157, 165

Anthropic principle, 88, 90, 91, 102, 163, 172, 176, 177

A priori/a posteriori, 37, 39, 40, 52, 60–61, 63, 91, 149, 150, 151–153, 154–155, 156, 157, 173, 237–238

Architecture, rectilinear shapes in, 210–214, 218

Aristotle, 22, 25, 26, 27, 39, 137, 146, 181, 191, 194, 247

Arms race, 193, 207, 230

Artificial selection, 220

Assimilation, 66, 67–68, 70, 72, 73, 114, 121. *See also* Accommodation; Adaptation

Association cortex, 103, 104, 108–109, 110, 111, 112, 123–125, 129, 131–132, 137, 139, 151–152, 242

Attachment, mother/child, 75, 77, 79, 98, 99

Austin, J. L., 1, 86, 213, 241, 249

Barnes, Jonathan, 194, 247

Barrow, John D., 90, 241

Bell, S. M., 77, 239, 240

Benjamin, Thomas, 227

Bennett, J. F., 238

Benowitz, L. I., 243

Berlin, Brent, 169, 246

Biology, 52, 53, 58, 59, 61, 93, 173, 182; genetic epistemology and, 64–65, 73; thoughts vs. things relationship and, 73–74; nature-nurture debate and, 93, 98, 191; environment as, 102; logic and, 148, 149; characteristics determined by, 179; evolutionary, 191. *See also* Neurobiology

Birdsong, 117, 119, 136

Blakemore, Colin, 121, 243

Blatt, S. J., 76, 140, 239

Bowlby, John, 78, 240

Brain (general discussion), 72, 97, 105, 164; experience of the environment, 97, 101, 115, 116–117, 119, 121; plasticity, 99, 106, 117, 118–119, 120–121, 122, 124, 127, 128, 242–248; understanding and, 107, 108, 109, 115; effect of injury and disease on, 108, 111, 115; activity within (relationship to itself), 131, 133; perceptual maps, 129–132, 135; values in, 139

Brain structure, 77, 79, 83, 94–96, 97, 101–102, 114, 123; evolution, 56–60, 61, 126–127; association cortex, 103, 104, 108–109, 110, 111, 112, 123–125, 129, 131–132, 137, 139, 151–152, 242; motor systems, 103–104, 106, 109, 110; surface, 104–105; basal ganglia, 105; cerebellum, 105; cerebral cortex, 105–107; limbic